DIGITAL SWITCHING
CONTROL ARCHITECTURES

For a complete list of the *Artech House Telecommunication Library*,
turn to the back of this book...

DIGITAL SWITCHING
CONTROL ARCHITECTURES

Giuseppe Fantauzzi
Telesoft S.p.A., Rome, Italy

Artech House
Boston•London

Library of Congress Cataloging-in-Publication Data

Fantauzzi, Giuseppe.
   Digital switching control architectures / Giuseppe Fantauzzi.
     p.   cm.
   Includes bibliographical references and index.
   ISBN 0-89006-452-0
   1. Telecommunications--Switching systems--Data processing.
  2. Digital communications. 3. Communications software. I. Title.
TK5103.8.F36   1990   90-43526
621.382--dc20   CIP

British Library Cataloguing in Publication Data

Fantauzzi, Giuseppe
   Digital switching control architectures.
   1. Electronic equipment. Switching circuits. Design
   I. Title
   621.381537

   ISBN 0-89006-452-0

© 1990 Artech House, Inc.
685 Canton Street
Norwood, MA 02062

International Standard Book Number: 0-89006-452-0
Library of Congress Catalog Card Number: 90-43526

10   9   8   7   6   5   4   3   2   1

To Chiara

Contents

Preface

Switching exchanges include so much software that it seems to grow endlessly. Features and characteristics of this software are sufficiently unique to make it a world apart in the universe of data processing. Hence, in this book we analyze the hardware and software architectures that are typically adopted for the control of telecommunication switching exchanges.

The book is the result of over 15 years of activities at ITALTEL, the leading Italian manufacturer of telecommunication systems. The material here has been used for the technical training of telecommunication experts, both at ITALTEL and in postgraduate courses. As such, the book can be useful as a reference or to teach university courses in digital switching systems.

This book is the outcome of long years of R&D activities, and the author is grateful to an almost endless list of people. However, very special thanks go to Mrs. Daniela Villani for her help during the early stages of this work, and to Mrs. Albertina Gambini for her patient support in typing the text and drawing the figures over and over again. Many thanks also go to Dennis Ricci of Artech House for his help in polishing the English of the book, which originally was by no means suitable for publication.

Giuseppe Fantauzzi
TELESOFT S.p.A. Rome
November 13, 1990

Chapter 1

Functional Structure of Digital Switching Exchanges

INTRODUCTION

Without referring directly to any specific system, the purpose of this chapter is to focus on *where* software and processors are typically used within switching exchanges. At the same time, as a further goal, similarities and differences are shown between normal data processing systems and the functional architecture of digital exchanges.

As a first step, a typical exchange is analyzed at a "black box" level to indicate which macrofunction it must perform to achieve the goals for which it is used within a public telecommunication network. The analysis of these functions permits us to show the similarities between exchanges and computers as well as their limits.

As a second step, a *canonical functional model* is introduced, which applies to each kind of exchange and focuses on a *quasibipolar* structure, consisting of a *common control* and a *switching matrix,* around which several kinds of *exchange terminations* are connected. Each different exchange termination is described, paying particular attention to the amount of *intelligence* that is intrinsically included and is therefore implemented by means of specific processors and related software.

After the exchange terminations, we consider the *switching matrix,* where the circuit switching of the calls occurs. Even in this case, the aim is to indicate software and processors that a matrix may also include.

In the concluding section, the main points are synthesized to identify *where* the software is located in a digital exchange. For the sake of brevity, the reader is expected to have a general knowledge of the realities and concepts typical of switching exchanges.

1.1 Switching Exchanges and Computers

A switching exchange is a technological system used in telecommunication networks to switch the calls arriving at its *terminations*. Today, there are basically two kinds of exchanges: *telephone* and *data* switches. Telephone exchanges typically switch *voice* and *other signals* (such as facsimile) that are physically equivalent to it. Conversely, data exchanges switch *data* or *telex*. However, with the development of *ISDN* (integrated services digital network), telephone exchanges are also becoming capable of switching data, video, *et cetera,* and therefore also tend to include the functions of data switches. Therefore, as will be shown later, digital switches and digital transmission treat voice and data alike, as simply bit streams.

For this reason, we will consider only the case of telephone switches, which are by far the most common and, in the near future, with the development of ISDN, will become even more pervasive than they are today.

A telephone exchange shown as a black box (see Figures 1.1a and 1.1b), interfaces with its external environment by means of a variety of *lines*. First in order of importance among these lines are the *analog subscriber loops* used to connect to the exchange *subscriber terminals* such as telephone sets, facsimile machines, *et cetera*. Within or outside of the subscriber loops, there may be analog *trunks* used to forward calls from one exchange to another.

A subscriber loop carries both the *useful signal* to be switched, such as voice, and the necessary *signaling* between the exchange and subscriber set. A trunk line carries the useful signal, which also includes signaling in the case of *associated signaling* trunks. However, *common channel signaling* trunks do not carry any signaling because this is sent by means of *data messages,* over *common signaling channels,* which are another kind of line in an exchange.

In order to maintain and operate an exchange, it needs to be provided with data lines that are different from those for common channel signaling and are used to connect the exchange to either data terminals or remote central computers. Additional data lines are used in exchanges that have ISDN subscribers to forward packet-switched calls to packet-switched data networks (PSDNs).

Telephone operator positions must be provided to implement a semiautomatic traffic function, where the switching is carried out by the exchange under the control of telephone operators (see Figure 1.2). These operator positions typically consist of a data terminal connected to the exchange by a data line and by two voice lines without signaling (this is because all the signaling between operator position and exchange goes through the operator data terminal).

Most of the subscriber lines carry analog voice and signaling. The same is true for analog junction lines. However, ISDN subscriber lines are fully digital. They can be of two types, called *base access* and *primary access*. A base access carries two *B channels* and one *D channel* on the same subscriber loop. The B channels operate at 64 kbit/s, the D channel at 16 kbit/s. The three channels are interleaved in a *time division mode* on the same two wire loop, and form a bidirectional data stream of $64 + 64 + 16 = 144$ kbit/s.

A primary access has two different standards called respectively $30B + D$ and $23B + D$. The 30B + D standard is the one mostly used worldwide, while

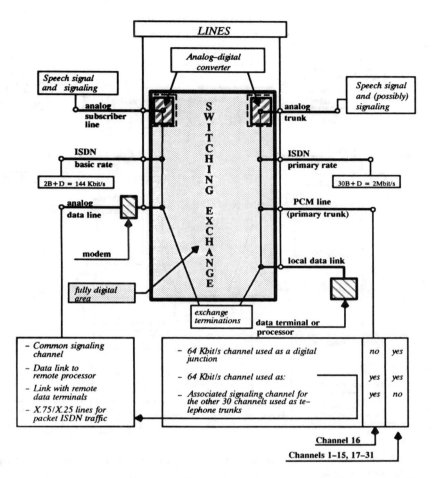

Figure 1.1(a) General layout of the interconnection lines for a digital switching exchange (European standard).

the 23B + D standard is used in North America. In both cases, each B channel and the D channel operate bidirectionally at 64 kbit/s. A 30B + D primary access forms a 2.048 Mbit/s (pulse coded) PCM line (European standard); a 23B + D primary access forms a 1.544 Mbit/s PCM line (North America standard). Note that what we call the European standard is, in fact, the international standard used in South America, Africa, Asia and the Pacific.

The B channels carry digital information such as digitized voice, data, digital facsimile, *et cetera*. However, the D channel of each access (either primary or base) carries the signaling related to the B channels belonging to the same access.

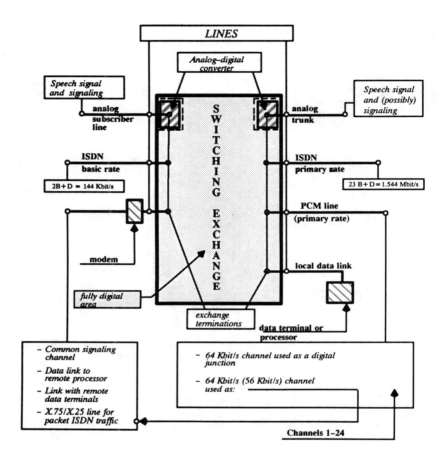

Figure 1.1(b) General layout of the interconnection lines for a digital switching exchange (North American standard).

For base accesses, the D channel may also carry data to be switched in packet mode. All this is done according to the relevant CCITT Recommendations.

Digital junction lines are multiplexed in time division mode over four-wire circuits and according to the CCITT Recommendations on PCM. Two classes of standards have been defined on this subject. The first is adopted by the European countries, the second in North America. In the European standard the basic multiplexing level operates at 2.048 Mbit/s (often, for the sake of brevity, instead of saying 2.048 Mbit/s, the European standard basic bit rate is referred to as 2 Mbit/s) and includes 30 + 2 bidirectional channels, each operating at 64 kbit/s. One channel, named *channel 0* is used for synchronization purposes. Another channel, named *channel 16*, may be used for signaling purposes. The other 30

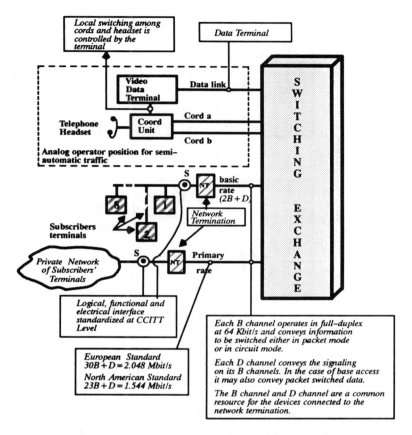

Figure 1.2 Connection layout for (analog) operator positions for semiautomatic traffic and for ISDN access.

channels typically carry voice digitized at a rate of 8000 byte/s. Any voice channel can also be used to transmit a data flow of 64 kbit/s. Channel 16, when not used to carry the signaling related to the other 30 voice channels, can be employed to carry a data channel operating at 64 kbit/s.

Several 2.048 Mbit/s lines can be further multiplexed in a time division mode, according to hierarchical levels normalized (standardized) by the CCITT. The first of these levels includes four 2.048 Mbit/s lines and operates at about 8 Mbit/s. These 8 Mbit/s lines may also be directly connected to exchange terminations. However, for the case of higher hierarchical lines (such as 34 Mbit/s) use is made of multiplexes which divide them into 2 Mbit/s lines, which are then connected to the exchanges (see Figure 1.3). The common signaling channels can be imple-mented by using *voice channels* of the PCM lines, or, in the 2 Mbit/s lines, in

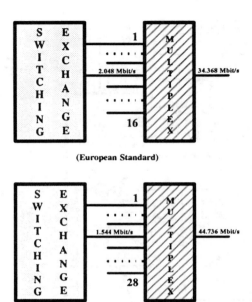

(European Standard)

(North American Standard)

Figure 1.3 Multiplexing-demultiplexing of PCM lines.

channel 16. In this case, the 30 voice channels of the same PCM line cannot be junctions with associated signaling. This concludes our review of the European standard.

In the North American standard, the basic PCM line contains *24 channels* and operates at a rate of 1.544 Mbit/s. Due to the approach used in this standard for synchronization and associated signaling, all 24 channels can be used as voice channels. Each can also be replaced by data channels, which may operate either at 64 kbit/s or 56 kbit/s (for more detail on this point, see [1]). Any channel in a 1.544 Mbit/s line can be used as a common signaling channel.

Except for the case of analog subscriber loops and analog trunks, all other lines carry digital information in the form of bit streams. Moreover, the signals related to the analog lines are converted into digital form by means of devices included in the exchange terminations. This means that, aside from these devices, *the entire digital exchange manipulates only data streams.* Its behavior, in fact, is that of a machine which processes incoming bit streams and as a result of its activity, provides outgoing bit streams. Both incoming and outgoing bit streams are carried by the lines connected to the exchange.

From a conceptual standpoint, such behavior could be emulated by means of a single computer with suitable software. Such a possibility, however, is only of theoretical interest. This is so for three reasons. First, the processing speed necessary for a computer to emulate an actual switching exchange with a few thousand lines is substantially above the capabilities of the technologies of today and those foreseen in the future at reasonable prices. For example, 2000 subscriber lines with 15 switched Erlangs for each 100 subscribers imply the need to transfer $20 \times 15 \times 2 \times 8.000 = 4,800$ Mbytes/s with a rate of 1 byte from line to line every 125 µs. Evidently, such a goal can be achieved only by computers with a processing capability of several megainstructions per second.

A second reason is that the reliability levels required in the exchanges entail the use of redundant structures, which can be implemented only by connecting two or more computers between them by means of a specialized network that cannot be considered as standard features of normal computers.

A third reason is that other data processing architectures, specifically designed for the exchanges, are available, and are much more cost-effective than the use of an individual standard computer.

Also, the use of standard computer networks, such as those considered in the theory and practice of computer science, are not competitive in this respect. As will be seen from the considerations developed in the following sections and chapters, successful approaches in the design of digital exchanges allow them to appear as specialized networks of several (hundreds, even thousands of) processors with typically nonuniform characteristics and properties. These processors are interconnected by a sort of hardware *cytoplasm,* implemented according to the techniques of sequential and combinatorial logic. This cytoplasm is a kind of intelligent interface, which not only interconnects the processors, but also carries substantial processing capacity. We can observe that the number of processors included in the exchanges tends to grow continuously, while the cytoplasm is continuously reduced, but without becoming marginal (see Figure 1.4). The tendency of the cytoplasm to remain viable is favored by the development of VLSI (very large scale integration) technologies, which are making hardware implementations of data processing functions more competitive with the use of software tools in which the same functions are implemented by means of microprocessors.

To conclude, the practical approaches which are actually followed in the switching exchange architectures have succeeded because they guarantee a higher level of flexibility with cost substantially proportional to the size of each exchange. At the same time, the exchanges also ensure better survivability when faults and outages occur.

1.2 A *Canonical* Functional Model for the Switching Exchanges

From the viewpoint of the functions that it performs, we can conveniently think about a switching exchange according to a quasibipolar reference model, where the two poles are a *switching matrix* and a *common control* with a multiplicity of *exchange terminations* and other kinds of units around them (see Figure 1.5).

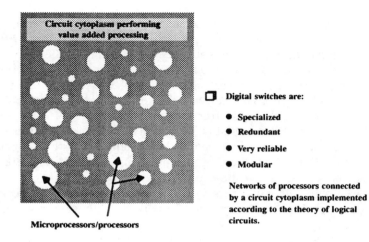

Figure 1.4 Switching exchanges and processors.

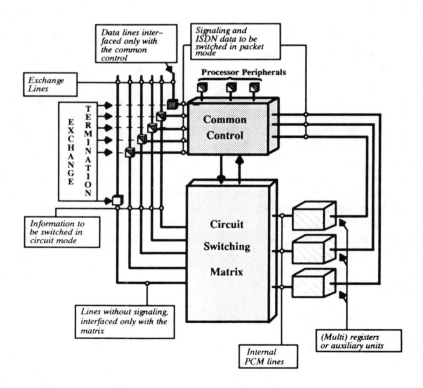

Figure 1.5 Layout of the quasibipolar model for telephone exchanges.

The *switching matrix* includes the functions by which the exchange performs *circuit switching* activities to carry out the physical, bidirectional, and uninterrupted transport between the lines involved (in each call), of the 8000 bytes/s of data or digitized speech.

The *common control* includes most of the processing control functions that take place in an exchange, and receives, processes, and forwards the signaling. Moreover, the common control specifies to the matrix which lines to switch with each other, when to start and when to terminate. In exchanges that can handle ISDN traffic, the common control carries out also the *packet-switching* functions. (For circuit *versus* packet switching, see Section 1.4.)

In the quasibipolar model, an *exchange termination* is an access point for a line, either analog or digital, to the exchange. The first function of a typical exchange termination is to separate the two components of the information stream flowing into a line: (1) *the useful signal to be switched in a circuit switching mode*, and (2) *the signaling or the information to be switched in packet mode*. The useful signal is transmitted through the matrix and the received signal is extracted; conversely, signaling and packet data pass from common control to the terminations. In some cases, either component can be missing. For instance, in the case of common signaling channel, there is no voice signal to be switched in circuit mode; however, a trunk line using common channel signaling does not carry any signaling. The characteristics of each exchange termination strongly depend on its lines. In the following section, we synthesize the most important cases on this topic.

In the functional model of the exchange, in addition to its matrix, common control, and terminations, we need to specify *in-band signaling registers, data multiregisters,* and *auxiliary units* (see Figure 1.6).

In-band signaling registers include the processing functions related to the multifrequency signaling that is carried on some types of trunk lines and on the subscriber loops that terminate at telephone sets with a multitone dialing device. In these cases, part of the relevant signaling is carried by combinations of voice-band tones. This is the usual medium of the voice signal, and it is typically used when such a signal is not being sent on the line. Each time a line is carrying multifrequency information, it is switched on a *register,* which converts the multifrequency information into messages for the common control. Conversely, the same register converts the messages that it receives from the common control into multifrequency signals, which are carried to the line connected to that register via the matrix.

In switching matrices, a more convenient way is to treat primary PCM lines (i.e., 2.048 Mbit/s lines in the European standard and 1.544 Mbit/s PCM lines in the North American standard), instead of basic 64 kbit/s channels. As a consequence, multifrequency registers are typically organized into *multiregister units,* consisting of a number of elementary registers equal to the number of channels specified for a PCM primary access. (In the case of the European standard, the

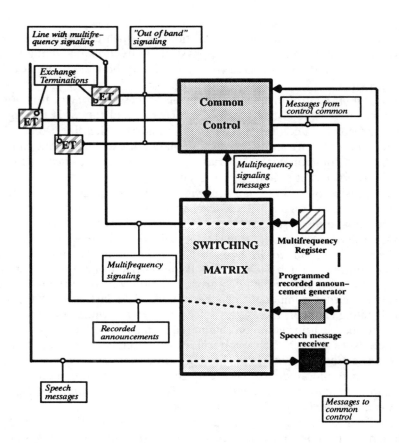

Figure 1.6 Connection layout of multifrequency registers, programmed recorded announcement generators, and speech message receivers.

number of elementary registers can grow to 32 because, within an internal PCM line connected to a matrix, channels 0 and 16 do not need to be used for synchronization and out-of-band signaling. In the North American standard, the number of registers can grow to 24.) The elementary registers belonging to the same unit communicate with the common control via a common interface, where we can concentrate some processing capabilities. This distributed processing allows for better use of the common control's internal capabilities and greater flexibility in the supervision of the actual operating mode of each elementary register and their common connection to the switching matrix (see Figure 1.7).

Note: Here, a first instance arises of a rather common situation where each unit or termination is connected to the common control via a device which has local processing capabilities that vary from case to case. Strictly speaking, such

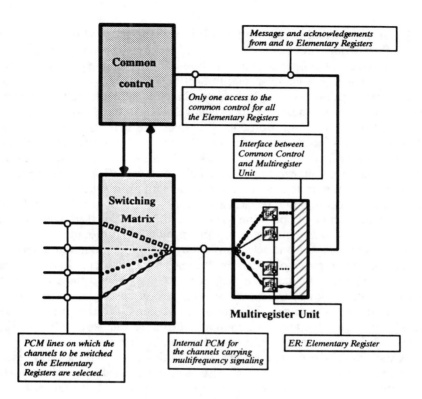

Figure 1.7 Multiregister unit: interconnection layout.

capabilities, at least from a theoretical point of view, can be considered part of the common control. However, our approach to practical exchanges makes this abstraction inappropriate.

An interface between common control and multiregister with a minimal processing capability performs the following functions:

- it receives signaling messages from common control, then translates and sends them to the chosen multiregister;
- it sends messages to common control, stating the occurrence of multifrequency combinations received at each elementary register.

With a higher level of processing, the same interface can execute testing procedures, either by itself or under direction of the common control. The results of these procedures are sent from multiregister to common control. Thus, we conclude our discussion of multifrequency registers.

Data lines and *common signaling channels,* connected to common control, must terminate at logical interfaces, allowing the common control to see each line or channel as a virtual circuit with which it exchanges *useful messages* instead of *uninterrupted sequences of bits* characterized by a given average error rate. The characteristics of these interfaces are analyzed, for the analog case (i.e., voice lines on which data are sent via modems) in Section 1.7. However, the most common case in digital exchanges is to have data channels implemented by means of 64 kbit/s PCM channels. Therefore, an exchange needs to have a specific kind of *data multiregister* for each data protocol with which the common control must deal. Such multiregisters interface with the matrix, similarly to the case of multifrequency registers, over primary PCM access lines and consist of a number of elementary registers equal in number to the channels of a primary PCM line (see Figure 1.7). Elementary registers in the same multiregister unit are connected to the common control through a common interface. Each time a given channel c of the PCM line is used as a data channel operating according to a protocol Π, c is switched by the matrix in a *semipermanent* way (i.e., for an indefinite amount of time until the connection is interrupted and reestablished by operation and maintenance actions) to an elementary register of a data multiregister for protocol Π. In this way, the elementary register will make the data channel c appear to the common control as a virtual entity with which it can exchange useful messages, instead of uninterrupted bidirectional streams of 64 kbit/s. The communication processes between each data multiregister and the common control are the same as those discussed in Section 1.7 for the case of lines for analog data channels.

A data multiregister is a rather complex device, which includes several microprocessors. These give to each multiregister the needed processing capabilities. The multiplicity of data protocols is a consequence of the fact that, at least in the most general case, an exchange must deal with:

- common channel signaling languages CCITT 6 and CCITT 7,
- X.75 lines for interconnecting data packet networks as needed for ISDN lines,
- X.25 lines for possible interconnections between the exchange common control and remote processing centers,
- the D channels at 64 kbit/s, part of the primary ISDN accesses.

Due to the differences among data protocols related to these cases, specific kinds of multiregisters must be used. There are actually two different but equally valid schemes:

- each register can be provided with needed processing capabilities to support a multiplicity of protocols, or
- a multiplicity of register types can be employed, each supporting a specific protocol.

In addition to in-band signaling registers and data multiregisters, an exchange typically may include auxiliary units, such as *programmed recorded announcement generators,* that are usually connected via the matrix on any line to which they must send a recorded speech message. Each time they are needed, the generators are first connected via the matrix to the appropriate lines by the common control, which then directs generation of the speech message. In the future, there could also be *speech message receivers,* which connected on a speech line, would allow for a subscriber to state verbally, digits and signaling addresses that would be recognized and converted into data messages for the common control. Speech message generators (not to mention speech receivers!) are rather complex units, implemented by means of intensive use of microprocessors and VLSI components. For historical and technological reasons, while registers are considered as conceptual extensions of common control, the other units mentioned here rather appear as *peripheral units,* specifically separated from it.

1.3 Exchange Terminations for Analog Subscriber Lines (B.O.R.S.C.H.T.)

A general layout for an exchange termination used for analog subscriber loops is shown in the upper part of Figure 1.8. (It comprises battery feed, overvoltage protection, ringing current, supervision, conversion, hybrid, and testing; hence, we have the initials, B.O.R.S.C.H.T.)

The first function that such a termination must give is the *battery feed* of the subscriber set connected to the remote termination of the loop, providing the loop current, which is modulated by the analog (voice) signals. Then, the same unit must give *overvoltage protection* for the exchange devices against meteorological, environmental, and industrial hazards. Each line termination is provided with a relay function to actuate the *ringing current* to the remote telephone set.

The other functions implemented by the terminations are (1) the *supervision* of the loop with the flow of signaling information between common control and the remote telephone (2) the digital-to-analog (D/A) and analog-to-digital (A/D) *conversion* of the speech signal, and (3) the telephone *hybrid* necessary to convert the two-wire signals available on the loop into four-wire signals, required in digital switches. As a final function, terminations must allow for extensive *testing* capabilities on the loop to detect the occurrence of problems such as foreign potentials and loop leakage.

The actual circuit layout used for the subscriber exchange terminations varies from case to case; therefore, Figure 1.8 should be used only as an example for easy reference in the considerations that follow. Starting from the subscriber loop, there is a primary protection block against overvoltages. The binary signals, *tip* and *ring* on the subscriber loop, are set by the common control and connected to switches. The terms "tip" and "ring" normally refer to the two copper conductors in the loop, with the negative potential (-48 volts) being applied to the tip and ground being applied to the ring. In a balanced loop, both tip and ring are switched and ringing current is applied across the tip and ring. The next block in Figure 1.8

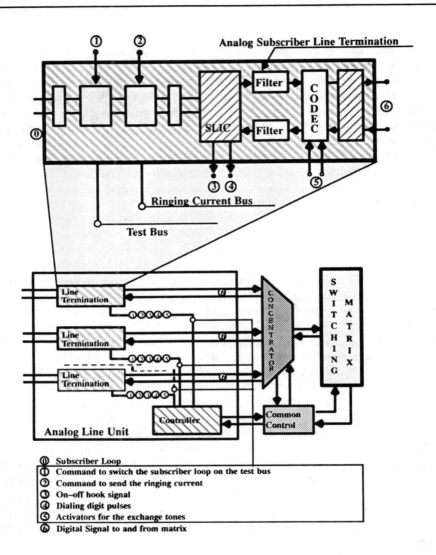

Figure 1.8 Analog line units and related exchange terminations.

is a secondary protection placed at the input of the kernel block of the termination, typically called SLIC (subscriber line interface circuit). It includes the power supply, subscriber signaling, and the two-to-four wire conversions. After the SLIC, one receiving and one transmitting filter are used to limit the signal bandwidth of the A/D converter. Such a converter, through a logical access, is connected to a concentrator used to merge more subscriber lines into a PCM line. Normally, the number of PCM channels available for the loops is smaller than the number of subscriber lines. This is to take into account the fact that, usually, only a fraction of subscribers communicate at the same time. The subscriber signaling, as already mentioned, is detected by the SLIC, which makes it available as a sequence of binary electrical signals to the common control. Binary signals are used by the common control to forward exchange tones to subscriber lines such as busy signal, dial tone, et cetera. A typical timing of the waveforms sent from SLIC to common control is shown in Figure 1.9.

During the called party selection phase, subscriber lines that use multitone telephone sets are switched on a subscriber in-band signaling register capable of detecting, digit by digit, for the common control, the pair of frequencies of the

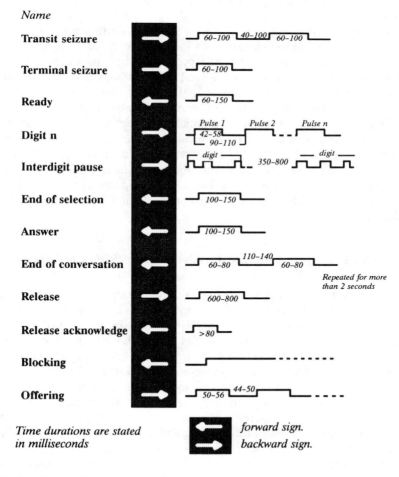

Figure 1.9 Decadic signaling over E and M.

called party generated by the calling telephone set each time a multifrequency tone button is pressed. The sectioning of the subscriber lines makes possible the selective connection of loop lines and exchange termination access points to auxiliary devices driven by the common control and used to perform the necessary test and maintenance functions. All this is implemented according to approaches that vary from system to system (see Section 1.12).

For each exchange termination, the common control must be capable of correctly decoding the binary signaling sent by the SLIC. Consequently, common control must actuate the binary signals to the termination in a way consistent with the information flow protocols defined for that unit. Finally, common control must carry out the operation and maintenance functions of the entire unit.

Subscriber line terminations are usually grouped into units, each having a common interface with the common control. Each subscriber line unit operates as a peripheral of the common control with message flows, as in computer applications. Subscriber line units are typically provided with internal processing capabilities related to the complexity of the messages sent back and forth between unit and common control. The processing capability of a unit differs from system to system. At the lowest level, the line unit must be capable of scanning the incoming binary signals from each exchange termination to detect every level change on each of them. Each change is communicated to the common control with the address of the selected line and typically with the duration since its last change. In the opposite direction, the line unit accepts from the common control and executes requests to switch, either to one or to zero, any signaling line to any of its line terminations. The number of terminations in the same unit is a system-dependent choice.

1.4 Basic Rate Interface Exchange Terminations for ISDN Subscribers

At the remote termination of an analog subscriber loop, either a telephone set is connected or a device emulating its behavior such as a modem or a facsimile machine. On the subscriber site, an ISDN line ends up in a *network termination* (NT). (see Figure 1.2). Several, possibly not identical, terminals may be connected to the same NT to share the subscriber's resources, consisting of two B channels and one D channel. Each device connected to the same NT is provided with its own address through which it can be reached by any other subscriber.

Each device connected to an NT can request the exchange to execute a connection to be operated either in *circuit mode* or in *packet mode*. In the first case, by far the most common, the exchange must provide a path between calling and called party for a steady bidirectional stream of 1 byte every 125 μs.

Note: In the second case, calling and called party send data packets to the exchange, in an asynchronous mode, which the exchange must be able to store and forward from the sending to the receiving terminal. A circuit switching mode connection is basically a transparent channel made available to the subscribers that use it continuously. However, a packet mode connection is equivalent to the availability of a postman accepting packets from either party and delivering them to the other. The postman gathers the packets one by one, puts them into a queue, and delivers them one after the other. Speech and facsimile communication operates in the circuit-switching mode. Data communication can operate in either mode.

The D channel first conveys the signaling related to every terminal connected to its NT. The same channel can also be used as a common link for packet mode communication with any terminal of its NT, whereas *each B channel may convey communication switched in either circuit or packet mode*. As a general rule, the terminals connected to an NT use their common D channel for the signaling process and for packet mode communication. Each B channel is a resource that a terminal connected to the NT seizes and occupies exclusively for the duration of a call.

ISDN subscriber lines are structured into *ISDN basic rate units* (see Figure 1.10), each including a system-dependent number R of terminations. Each basic rate ISDN unit interfaces the network by means of R subscriber loops. On the matrix side, the same unit has $2 \times R$ four-wire circuits, each conveying a different B channel. On the common control side, a communication channel is used both for the signaling and the data packets conveyed on the D channels of the unit. From a functional point of view, a basic rate ISDN unit consists of a *line termination* (LT), repeated R times, plus a centralized unit that is the interface with the common control. Each LT executes the following functions:

- Transduction from the electrical levels of the loop to the logical levels used in the exchange and *vice versa*;
- Two-to-four wire conversion;
- Demultiplexing-multiplexing of the two B channels and the D channel from and to the 144 kbit/s bidirectional stream used in each ISDN line.

The B channels of each LT are multiplexed and forwarded on internal PCM lines. However, the D channels must be locally processed in a way which is substantially similar to that of the common signaling channels (see Section 1.7). Such processing, in the incoming direction from loop to exchange, includes the reception of the frames transmitted by the terminals on the D channel and the implementation of line protocols. This identifies faulty frames and requests their retransmission. In the outgoing direction, one needs to implement the activities corresponding to those executed in the receiving direction. The LTs in the same unit communicate with the common control via a common interface with processing capabilities that vary from system to system. Such an interface, at least, sends the common control all incoming messages and packets from the LTs, and distributes messages and packets originated by the common control for its units' D channels. Typically added

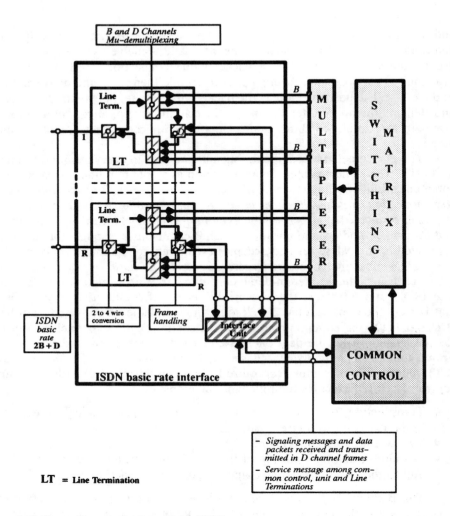

Figure 1.10 General layout of a line unit for ISDN access.

to this kernel of basic functions are operation and maintenance functions for the entire unit. Such functions are carried out under the supervision of the common control, which sends the directives and receives the subsequent responses.

1.5 Exchange Terminations for Analog Trunks

An analog trunk is seen by its exchange termination as a set of physical wires carrying voice and signaling. As a particular case, trunks with common channel signaling carry voice only signals. An exchange termination for an analog trunk (see Figure 1.11) first consists of a *trunk line* that separates voice from signaling, conveying the former to the matrix, the latter to the common control. From the signaling viewpoint, a trunk interfaces the exchange in a standard means by way of one or two binary signaling leads E from trunk to common control and with the same number of M leads from common control to trunk. The actual number of E and M leads depends on the signaling language of the junction line. The signaling leads E and M convey rectangular binary waveforms such as those shown in Figure 1.9.

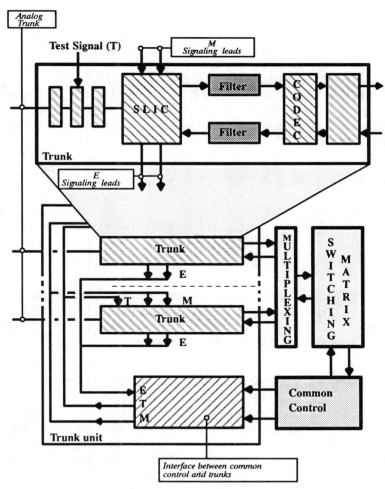

Figure 1.11 Trunk unit.

Multifrequency trunk lines, in addition to the signaling conveyed by *E* and *M* also make use of multifrequency (*in-band*) signals. These trunks are connected via the matrix to suitable registers, which interpret the in-band signals and pass translated data messages to or from common control, then inject the signals into the lines.

Also, the trunk lines are organized into units, often called *trunk groups,* with characteristics substantially similar to those already observed for the analog subscriber line units.

1.6 PCM Line Exchange Terminations

An exchange termination for a PCM line interfaces all the junction lines multiplexed on the same line. The functions of an exchange termination are rather complex and briefly described in the following paragraphs by referring to the diagram of Figure 1.12. Hereafter, specific reference is made to the case of 2.048 Mbit/s PCM lines. However, the case of 8 Mbit/s lines and that of North American standard lines are substantially similar if we ignore elements of secondary importance for our purposes.

A. Incoming Line Electrical Interface

This block physically interfaces the line with the exchange. Its purpose is to restore the incoming signal after distortion caused by the transmission network.

B. Line Clock Synchronization

To read the incoming bit stream the line must be synchronized to the clock of the incoming line. Only in this way, in fact, can ones and zeros be read in the incoming signal when the probability of incorrect wrong choices is lowest. This function is carried out by techniques that vary from system to system.

C. Thresholding and Interpretation—Bipolar-Unipolar Conversion

At this stage, the incoming signal (i.e., a distorted, noisy, analog signal) is interpreted, and, via a thresholding device, a string of logical (perfectly shaped) ones and zeros is generated for use by the switching matrix. This regeneration of the bit stream is what makes digital transmission and switching superior to their analog counterparts.

Note: The restored incoming signal must be converted from the bipolar *line code,* such as AMI or HDB3 (see [1]), into the unipolar code suitable for processing

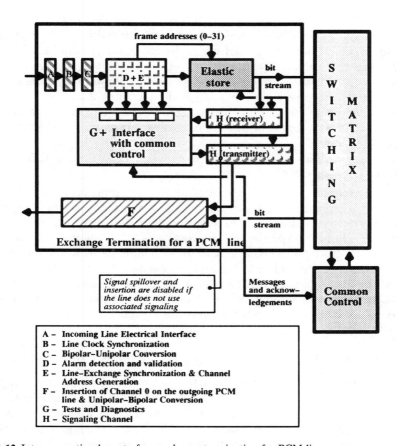

Figure 1.12 Interconnection layout of an exchange termination for PCM lines.

in the exchanges. This function is carried out by means of *ad hoc* LSI (large scale integration) circuits. As is well known, AMI and HDB3 codes provide for an electrical signal that is suitable for transmission on the line. We use codes because the normal sequence of zero and one levels, typically used in the electronic components of switching exchanges, is not adequate for actual transmission on physical lines.

D. Alarm Detection and Validation

After a signal has been converted, the occurrence of possible anomalies on the transmission line must be detected through the actual reading of the incoming bit stream. Anomalous situations are typically encoded by means of alarms such as the following:

- *Loss of frame synchronization*—Loss of synchronization on the frames, as a result of which the unit cannot correctly detect the stream related to each individual trunk line.
- *Lack of pulses in the receiving direction*—As the definition says, this condition is detected each time a bit is missing in the incoming direction.
- *Error rate above threshold*—This condition occurs each time an error rate above 0.001 is detected on the frame synchronization words A and B (see following point E).
- *Termination alarm*—Termination alarm is set each time a failure on the exchange termination is detected.

Alarm signals such as those described above are preprocessed and filtered in the exchange termination, which communicates only steady alarm conditions to the common control. This is done to avoid random alarms, which may be due to noise on the PCM line. Typically, the processing of the elementary alarms results in two kinds of information being sent to the common control: one concerns line failures; the other relates to line termination problems. Of course, common control could perform this filtering function just as effectively. However, the important point is that this is an example of how a multiplicity of peripheral processors or microprocessors can perform routine functions to reduce the central processing load.

E. Line-Exchange Synchronization and Channel Address Generation

To switch correctly, the exchange must be synchronized to the incoming PCM frames to be able to recognize the time slot of each channel multiplexed on the same PCM line. For this reason, a 30 + 2 PCM line has a channel 0 in which, two *A* and *B* patterns, internationally normalized, are used as synchronization flags. The exchange termination must be able to detect these patterns to synchronize to the incoming line. In this way, the same unit will be able to detect the position of each PCM channel within the incoming bit stream. The unit will then send channel address and sample value for each time slot to an *elastic store,* once every 125 μs for each channel. The elastic store is a device which allows for the synchronization between incoming bit stream and internal exchange clock. It is a memory with two pointers: one to write at the speed of the incoming PCM line stream, the other to read at the rate of the exchange clock. The occurrence of possible overlapping of the two pointers causes either the loss or the repetition of a frame arriving from line to exchange. These events are notified to the block *G,* which deals with alarm management.

F. Insertion of Channel 0 on the Outgoing PCM Line and Unipolar-Bipolar Conversion.

Before being transmitted, the unipolar PCM signal coming from the matrix must be organized into continuous PCM frames. For that purpose, a specific circuit alternately inserts synchronization patterns A and B in channel 0. Another circuit converts the final PCM bit stream in the transmission code (e.g., AMI or HDB3) relative to the line being used. As the final step, the physical interconnection with the trunk is made.

G. Tests and Diagnostics

For reliability reasons, exchange terminations for PCM lines are typically duplicated (i.e., redundancy), starting from the elastic store. A failure located on one section only causes the isolation of that section and causes the other to operate as the master. At the same time, the faulty section is diagnosed to see where and how it needs to be repaired. When both sections become faulty, the whole exchange termination is taken out of service. A PCM termination, due to its complexity, must be capable of testing to ensure that it is operating properly. So, loopbacks can be set in each termination so as to isolate exchange from lines and to allow for testing procedures by the common control. Tests executed on only one section of the exchange termination are periodically scheduled, temporarily taking that section out of service.

H. Signaling Channel

Note: Hereafter the signaling problem is treated only for the case of the CCITT international or European standard. In this respect, the North American approach is quite different [1]. However, its implications with regard to processing capabilities and software in a PCM termination do not vary greatly. Therefore, the example of the European case should suffice.

A PCM line transports trunks, which may carry either data or voice. When used for normal telephone calls, they may bring associated signaling by means of E and M leads. In analog trunks, the signaling leads are binary variables conveying continuous rectangular waveforms used for telephone coding. When a junction is multiplexed on a PCM line, these waveforms are also sampled and their values are placed in channel 16. To make this possible, the concept of *superframe* (see Figure 1.13) is defined as a sequence of 16 consecutive frames labeled from 0 to 15 and repeated without interruption at a rate of a superframe each 2 ms (2 ms = 125 μs $\times$ 16). The frames within the superframe differ from each other only in the time slots of channel 16. In each superframe, channel 16 of the first frame always contains the pattern 000001XS, where X is set at 1 when the originator of the frame wants to signal back that it is momentarily incapable of receiving superframes. S is a bit that, being available at the beginning of each superframe (i.e., once every 2 ms), may be used as a data channel operating at 500 bit/s. The pattern 000001XS is used in the receiving direction to acquire superframe synchronization and to know at each moment which frame of a superframe is being

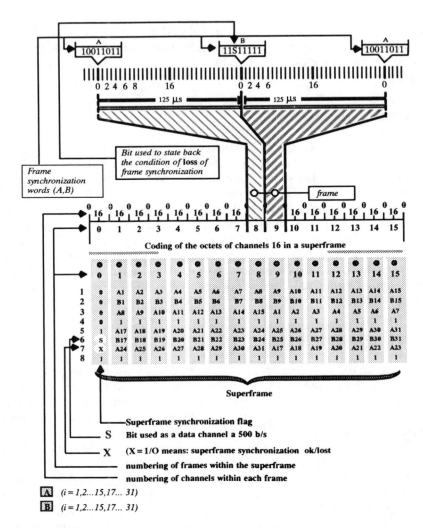

Figure 1.13 PCM frames and superframes (European standard).

received. The situation depicted in Figure 1.13 shows how the coding of channel 16 is defined in each frame T_i. In the Figure A_i stands for the first signaling lead of the *useful channel i* ($i = 1, 2, \ldots, 15, 17, 18, \ldots, 31$) in the PCM line. B_i denotes the samples of a possible second signaling lead of the same channel. Each A_i is read twice in each superframe or, equivalently, once every 1 ms. Each B_i is sampled once in each superframe. Based on the situation shown in Figure 1.13, a PCM line exchange termination in the incoming direction is able to recognize the E incoming leads of the 30 useful channels. At the same time, in the outgoing direction, the exchange termination can sample the M leads and put them in channel 16 of the relevant frames. All of this is done for one superframe after another.

As should be evident from our previous considerations, a digital line exchange termination is a rather complex device, which must interface the common control to transceive, as needed, the telephone signaling of trunks provided with associated line signaling.

Moreover, between termination and common control, there must be a flow of messages to carry out the supervision and maintenance functions for the entire termination. Similarly to the case of exchange termination units for analog junctions, the processing capabilities associated with each termination vary from system to system. A PCM termination unit must at least be provided with the capability of communicating to the common control the changes of state of the signaling leads from the trunks. Each variation will specify the new value of the lead, the time spent in the previous value, and the lead identifier. In the opposite direction, the minimum processing capability of a PCM unit includes its ability to accept messages from common control, and to implement them on the lines, such as: *set at 1/0 the outgoing signaling lead M1/M2 of channel T*.

The increasing levels of processing capabilities include, first, the ability of the unit to identify complete telephone criteria from the samples of the incoming signaling leads of each junction. Conversely, the unit may accept more complex commands from common control, such as the actuation of complex telephone signaling on the outgoing leads. Moreover, the unit can be given substantial capabilities to carry out its own supervision and maintenance activities.

Note: In any case (see Figure 1.14), as already noted, there are situations in which trunks multiplexed in PCM lines *do not* include associated signaling and therefore *do not* make use of channel 16. When this happens, the related functions of the exchange terminations are inhibited.

1.7   Exchange Terminations for Common Signaling Channels

Note: The channel signaling languages commonly used and normalized as international standards are CCITT 6 and CCITT 7. Although their differences are wide and substantial, they have been built on a base of common concepts, typical of data transmission techniques, which determine the fundamental characteristics of the related exchange terminations. This makes possible the analysis of such exchange terminations by referring to a common substratum of CCITT 6 and CCITT 7. This approach is followed hereafter.

Both signaling languages use full-duplex, synchronous data channels (see Figure 1.15). This means that each exchange termination for common signaling channels sees on its incoming line an uninterrupted bit stream characterized by a uniform bit rate; in the opposite direction, at the same bit rate, a continuous

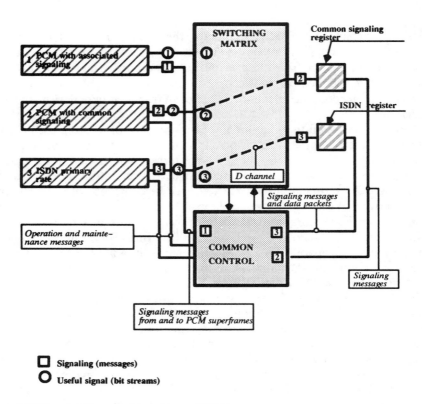

Figure 1.14 Signal spillover and insertion on PCM lines.

outgoing bit stream is generated. The bit streams on a common signaling channel are organized into consecutive *frames*. In the incoming direction, an exchange termination must be capable of synchronizing itself to the incoming stream at a *bit level*. Moreover, whenever such an activity is not successful, the exchange termination must communicate this event to the common control.

After being synchronized at a bit level, the exchange termination must acquire *frame synchronization* to be able to understand correctly the beginning, content, and end of each frame. The frame synchronization process is carried out according to procedures which may be different in the two CCITT languages, but both are based on the identification of specific repetitive patterns in the incoming bit stream. When synchronized at the frame level, an exchange termination analyzes each frame that it receives. Each frame is internally redundant and numbered. The internal redundancy of each frame (i.e., the insertion of some bits by which the frame can be checked) is used by the exchange termination to verify its correctness; the numbering is used to retransmit any frame which does not arrive correctly at

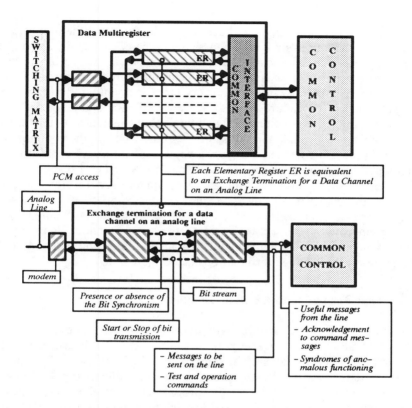

Figure 1.15 Exchange termination to handle data channels and common signaling channels.

its destination. Each frame that is not error-free is discarded by the receiving termination.

An exchange termination for a common signaling channel must carry the bit and frame synchronization functions. In so doing, *it transforms the common signaling channel, which appears to the common control as if conveying only useful messages.* These messages include those actually forwarded on the signaling channel and those which common control and terminations send to each other to carry out supervision, test, and diagnostic functions. The two terminations at the ends of a channel, during the course of their communication protocol activities, check one against the other to ensure that both are working properly. In this way, each termination can specify to its common control if, when, and in what sense anything is not working properly on the signaling channel. At the same time, a common control may begin test and diagnostic procedures for its terminations, for which it obtains the relevant responses.

Because of its complexity and the nature of the functions that it must execute, an exchange termination for a common signaling channel is built around one or more microprocessors. Moreover, some of its functions, such as those related to error control in the frames and frame synchronization procedures, tend to be carried out by means of *ad hoc* VLSI (very large scale integration) components, which have built-in procedures stated in the CCITT Recommendations. An exchange termination for a common signaling channel includes a substantial processing capability, which may, in this sense, be removed from the common control. The level of processing needed is a system-dependent choice. A common signaling channel may be implemented by either analog trunks or PCM lines.

In the second alternative, the PCM channels used for this purpose are switched via the matrix on specific *data multiregisters* compatible with the signaling language being used in each case. On the common signaling channels to which they are connected, these registers execute the same functions as in the case of common signaling channels implemented by analog lines and arriving at the exchange via specific exchange terminations (see Figure 1.15).

1.8 Exchange Terminations for Data Channels

A common control can be provided with mass memory devices that are connected to it, as is usually the case in computer architectures. This is also true for other kinds of standard peripherals such as printers, terminals, *et cetera*. Typically, there can also be data links, similar to the common signaling channels, but used to connect the common control to remote processors. Data channels in PCM lines use the switching matrix to make circuit-switching connections between registers and the physical lines. For each analog data link, a specific exchange termination is provided, similar to those used for common signaling channels. The differences in this approach lay in the details of the operating procedures on the lines, which are, in each case, functions of the data language being used.

1.9 ISDN Primary Rate Interface

As already mentioned, to connect ISDN subscribers with more than two B channels, in the European standard, a 30 B + D primary rate interface is available, consisting of a 2.048 Mbit/s PCM line, where the 30 channels (1 to 15 and 17 to 31) are the 30 B channels, and channel 16 is the D channel, working at 64 kbit/s and used for the signaling of the 30 B channels. In the North American standard, the primary access operates at 1.544 Mbit/s and includes 23 B channels and one D channel.

To deal with the ISDN primary rate interface, specific exchange terminations are provided, which are similar to those used for ordinary PCM lines. In this case, of course, there is no treatment of E and M signaling and the D channel is switched in a semipermanent (nailed-up) manner toward an *ad hoc* data multiregister for 64 kbit/s D channels. All the considerations related to characteristics and processing capabilities of the PCM exchange terminations apply as well for ISDN primary rate interface.

1.10   Circuit-Switching Matrix

Within a telephone exchange, its *packet-switching functions are implemented in the common control*. However, all circuit switching is carried out by the switching matrix; independently of its architecture, the switching matrix is not only an abstract function, but a subsystem specifically designed and optimized to perform circuit-switching functions.

A digital matrix, considered as a black box, interfaces the external environment by internal PCM lines, either primary rate (i.e., 2.048 or 1.544 Mbit/s) or others, operating at rates that are multiples of the primary ones. Analog lines, for both subscribers and trunks, are first digitized, then either concentrated or multiplexed on PCM lines. All of this occurs within the exchange termination units.

The circuit switching of two voice lines multiplexed on PCM lines is a rather complex operation. To focus on it briefly, the case depicted in Figure 1.16, in which switching channel 5 of line a with channel 21 of line b is required, can be considered as a meaningful example. We can assume, as is normally the case in actual exchanges, that the whole matrix and its lines are synchronized on the same bit and frame clock. This means that the time is divided into frame intervals lasting 125 μs each. Within each frame interval there are 32 time slots, labeled from 0 to 31. Starting and ending instants of each frame and of each time slot are identical within the matrix and for any incoming line.

Referring to such a situation, to switch channel 5 of line a with channel 21 of line b, means to execute, *once every 125 μs*, the following two operations:

- Removing from line $a,$ at time slot 5, the byte arriving at the exchange and forwarding it in the succeeding time slot 21 of the same frame on the outgoing PCM line b.
- Removing from line b at time slot 21 the byte arriving at the exchange and forwarding it at time slot 5 of the next frame interval over the outgoing link of a.

These functions can be efficiently carried out only by specific hardware architectures, which constitute a basic topic in both theory and practice of switching exchanges. This topic is not considered here because it is beyond the scope of our book and cannot be summarized in a few sentences. A very brief discussion of this point is provided, only for the sake of completeness, as an appendix to this chapter.

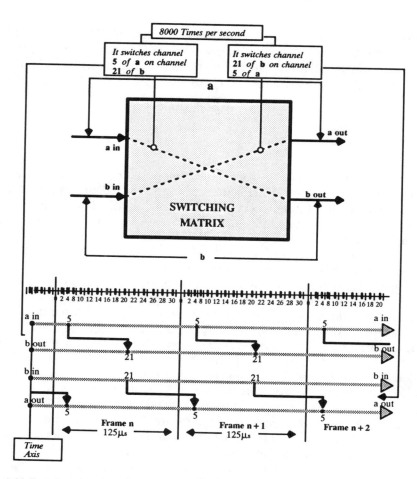

Figure 1.16 Functional principles of digital circuit switching.

A matrix, considered as a black box with several PCM lines, must interact with the common control, which gives it all the necessary commands to connect and disconnect the matrix PCM channels. In the opposite direction, there is a flow of responses from matrix to common control. As a fundamental part of the exchange, the matrix needs to be tested and diagnosed exhaustively, efficiently, and quickly. These maintenance functions are implemented by using hardware and firmware devices which are part of the actual matrix. Typically, these devices make extensive use of processors, so the functions that they implement, at least from a theoretical viewpoint, can be considered as a part of the common control. However, to approximate the technological realities of the exchange, a more convenient way to think of them is as partially belonging to the matrix proper, and partially included in an intelligent interface between common control and matrix, typically called a *marker* (see Figure 1.17). The implementation approaches of switching matrices in digital exchanges make extensive use of special VLSI components, which, together with microprocessors and related memories, constitute most of the matrix.

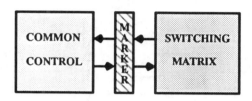

Figure 1.17 Markers within exchange.

1.11 Packet Switching

In a telephone exchange, the most often used switching technique is *circuit switching.* However, with the introduction of ISDN, telephone exchanges must be also capable of executing *packet switching,* for the following reasons (see Figures 1.18 and 1.19):

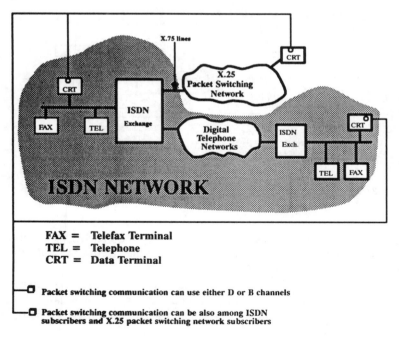

Figure 1.18 Packet switching in an ISDN network.

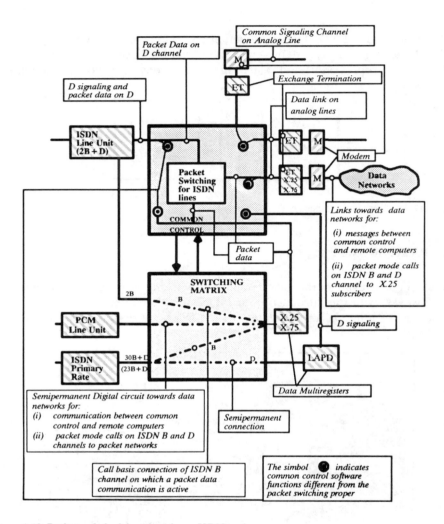

Figure 1.19 Packet-switched functions in an ISDN exchange.

- On the D channel, an ISDN subscriber may require X.25 packet-switching functions, with subscribers that can belong either to the same ISDN network or to X.25 packet-switched networks.
- On the D channel that is used as the signaling channel, a subscriber may need to start a packet-switched communication on a B channel with ISDN subscribers having the same services and with any subscriber belonging to a public X.25 packet-switched network.

Packet-switched functions are carried out by the common control. In this respect (Figure 1.19):

- The D channels of ISDN primary rate interfaces are semipermanently switched on data registers capable of handling them. This allows for the common control to exchange signaling messages with each 64 kbit/s D channel.
- Each B channel, while conveying data packets, is temporarily switched on a data register capable of handling X.25 frames. In this way, the common control can exchange data packet with each B channel conveying an X.25 communication.
- PCM channels, used as 64 kbit/s X.75 data links towards X.25 packet exchanges, are conveyed in a semipermanent way (i.e., "nailed up") on suitable data multiregisters. This allows the common control to exchange packets with each 64 kbit/s link.
- X.75 data links towards packet switched exchanges may also be implemented by means of *analog data links* terminating on specific exchange terminations.
- The D channels for ISDN basic rate interfaces convey packet data through exchange termination units. This approach allows the common control to treat these packets as necessary.

1.12 Testing Devices

In addition to the pervasive internal testing circuits, a switching exchange includes a set of testing devices (see Figure 1.20) used to check its lines and signaling systems. Each testing device is basically one that can be programmed to generate and receive calls to or from exchange terminations.

The testing device is also capable of step-by-step verification of the signaling processes and measurement of the transmission characteristics of the lines. Typically, a testing device is connected to one or more trunk lines and the common control. Usually, such a device can execute several different sequences; it executes those stated by the common control to which it sends back measurements and responses. Subscriber lines are provided with line unit testers, which (under the supervision of the common control) carry out transmission measurements and emulate signaling processes. These units are connected to access buses that cross all line terminations. Each line can be cut and connected to the access bus by the common control so that the line unit tester can check either the exchange termination or subscriber loop. In this way, the common control can start test and measurement routines on the lines. We will show in the following chapter that testing devices and line testers are operated by the maintenance software of the exchange. Typically, these units are complex devices, implemented by one or more processors and making use of software techniques. These devices have to be con-

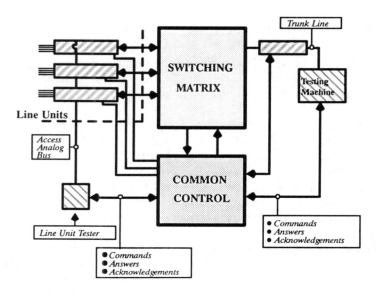

Figure 1.20 Testing device layout.

sidered autonomous peripherals of the exchange and not directly of its common control. The opposite viewpoint, in fact, even if it were acceptable theoretically, would be an unrealistic way to view practical digital switching exchanges.

1.13 Intelligent Networks and Additional Subscriber Services

In addition to normal telephone service, which basically allows every subscriber to call any other, with or without ISDN, switching exchanges offer several other special services, as illustrated in Figure 1.21. These services are implemented by software, which allows more complex telephone calling. These special, more complex calls may be carried out by either individual exchanges or an interaction between each exchange and special centralized databases. The need for centralized databases arises, for example, in the case of the so-called 800 service in the United States in which, by dialing 800 followed by the telephone number, the calling subscriber will reach a called party that in principle can be located anywhere on the network. (For example, you are calling the local number of a company at night and your call is switched to another number at that company where you can get an answer.) To know where such a call must be routed, each exchange involved will question the centralized database, which will answer with messages specifying all the relevant information required for putting through the call. Note that the databases to which we refer are not used for operation and maintenance functions, but rather for actual telephone traffic. Networks that make use of databases are

---

– *Incoming call barring.* It allows the subscriber to prevent all or certain incoming calls by dialing special sequences on the telephone keyboard

– *Outgoing call barring.* It allows the subscriber to restrict the area code to which calls will be sent from the telephone station

– *Timed hot line.* It provides the subscriber with automatic call origination to a predetermined number if within an administrable delay after going off hook, digit dialing is not started

– *Abbreviated dialing.* It provides a subscriber with the ability to program numbers to be dialed with an abbreviated code

– *Call waiting.* It alerts the subscriber talking on the line that another call has reached the telephone station. The called subscriber can switch between the calling parties

– *Call forwarding.* Incoming calls are forwarded to another telephone number, programmed at the telephone set.

Figure 1.21 Advanced service examples.

typically called *intelligent networks* to stress the fact that centralized processing capabilities are common network elements, and not attibutes of each individual exchange.

CONCLUSIONS

In a digital telephone exchange a distinction must be made between the useful digitized signal (i.e., the signal which must be switched) and the multiple forms of signaling used by the exchanges to switch telephone traffic and to carry out operation and maintenance functions. The circuit-switching activities are carried out by the matrix; the signaling and packet-switching are handled by the common control, either directly or indirectly via distributed processing.

Around the common control and matrix there are several kinds of exchange termination units, which interface the lines of the exchange with the common control or matrix. The interface functions available in the termination units may include variable levels of processing capabilities, which make them appear as more or less intelligent peripherals of the common control. Levels of processing vary for each type of termination unit, according to functional architectures that differ from system to system.

Also, the switching matrix includes substantial internal processing capabilities, which, however, appear more often as *firmware functions* (analytical control functions of hardware elements, implemented by programming processors integral to the matrix). Typically, the common control interfaces the matrix through markers, which are likewise intelligent units implemented by microprocessors.

At least from a conceptual viewpoint, processors and processing capabilities in the exchange termination units and markers may be seen as peripheral to the activities of the common control in an exchange. This is so, even if in actual architectures they may be implemented in a decentralized and distributed way.

In addition to common control, matrix, terminations, and markers, an exchange also includes several kinds of peripherals, such as multifrequency registers, data registers, testing devices, recorded announcement generators, *et cetera*. All these are peripheral units for which a wide variety of hardware and software techniques are used. In any case, registers should be considered natural extensions of the common control. However, testing devices, or recorded announcement generators rather appear to be external peripherals which may require connection to the common control.

APPENDIX: AN EXAMPLE OF SWITCHING MATRIX

For the sake of completeness, an example concerning the treatment of circuit switching in a digital matrix is shown here. The example considered here is based on a practical instance in which use is made of a VLSI component (available on the market), capable of switching, without loss, up to eight 2.048 Mbit/s PCM lines. The logical layout of such a component, called ECI from the Italian acronym for *elemento di commutazione integrato* (Integrated Switching Element) is shown in Figure 1.22. The ECI interfaces with the outside world by:

- eight outgoing PCM lines;
- eight incoming PCM lines;
- one communication channel with a controlling microprocessor;
- the usual synchronization and power supply signals.

The ECI is controlled by two external clock signals; one operating at $2048 \times 2 = 4096$ kHz, the other operating at 8 kHz.

The first clock is used to scan incoming and outgoing bits on a PCM line; the second is used to state the moment in which a frame is starting on each line. The ECI is structured (see Figure 1.23) around two memories: a *signal memory* (MS) and a *control memory* (MC), both having 256 cells of 8 useful bits. The MC memory also has one control bit for each cell.

The 256 cells of MS are structured into 8 consecutive groups with 32 cells each: one group for each incoming PCM line. In each group, its 32 cells are biunivocally corresponding to the $30 + 2$ channels of an incoming PCM line. Once every 125 μs, each cell is updated with the last incoming byte from the associated channel of the PCM line for that cell. At the same rate (i.e., 8000 times per second)

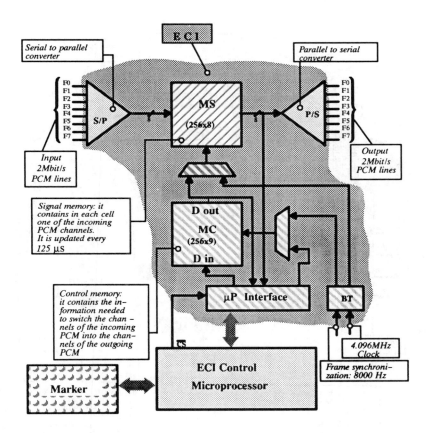

Figure 1.22 Integrated switching element (ECI) general layout.

each cell of MS is read and its content is transferred on an outgoing PCM line. The memory MC is used for that purpose. Its cells are in a one-to-one correspondence with the channels of the outgoing PCM lines; also MC is structured into 8 groups, each having 32 cells, one group per line. In each group, the 32 cells are associated on a one-to-one basis to the 30 + 2 channels of the outgoing line.

The typical MC cell associated with the output channel o_c ($o_c = 0, 1, \ldots,$ 31) of the outgoing PCM line o_l ($o_l = 0, 1, \ldots, 7$) contains on three bits the address i_l of an incoming PCM line, and on five bits the address i_c of a channel of that line.

The meaning associated with the MC cells addressed (o_c, o_l) is that every 125 μs, when its time slot arrives, the channel o_c of the outgoing line o_l is filled with the content of the MS cell at that moment having its address (i_c, i_l) written in the cell of MC at address (o_c, o_l), which means the last sample to arrive at MS from channel i_c of incoming PCM i_l. This process repeats for every channel of each

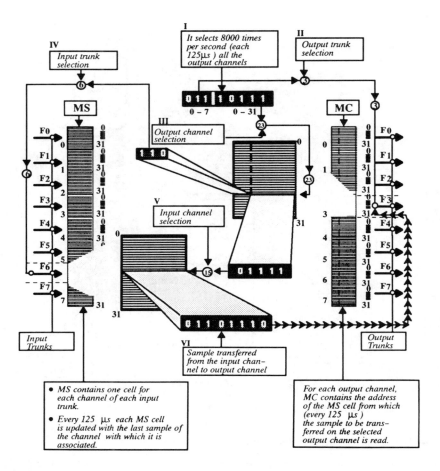

Figure 1.23 ECI: Functional layout.

outgoing PCM line. Thus, through the MS and MC memories, circuit switching is possible in every channel of any incoming PCM line to any channel of any outgoing PCM line. Moreover, each possible combination from incoming to outgoing channel is feasible. An ECI allows for all its channels to switch among each other without loss. Its MC memory has 9-bit cells because the 9th digit is necessary to implement control functions. In detail, for transfers from incoming lines (i.e., from MS) in MC, the bit is set at 0. Conversely, when it is set at 1, the 8 bits of the same cell are those which actually must be sent, once every 125 μs, to the outgoing channel associated with the cell (this feature is used to test the ECI). The microprocessor can control the ECI by updating its MC memory, cell by cell. The same microprocessor can read MC and MS memories. This, of course, is done for diagnostic reasons.

The limitation of the ECI is its capacity, which is 256 channels. If higher capacities are needed, as is normally the case, more ECIs must be connected to form a *switching matrix*. Structures and characteristics of these matrices are one of the main topics in the theory of switching exchanges. As an example, Figure 1.24 shows a particular network of ECIs. The marker between common control and the ECI network must select, for each circumstance, the optimal path from line to line and command the ECI processor to execute the chosen connections and disconnections. An ECI network may be designed to minimize the statistical probability of unsuccessfully routing a call from end to end. This is so, if not

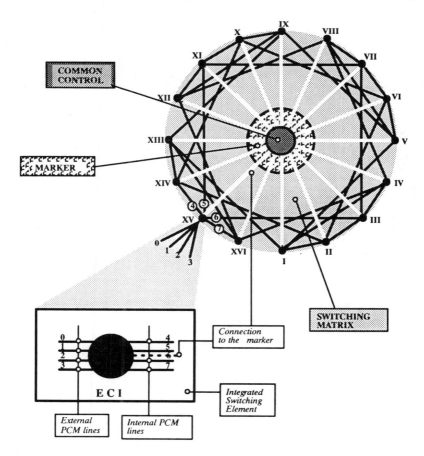

Figure 1.24 Digital switching matrix built as an ECI network.

always, at least as long as the traffic handled by the exchange remains within the limits for which it is engineered.

REFERENCES

1.   John C. McDonald, ed., *Fundamentals of Digital Switching,* Plenum Press, New York, 1983.
2.   GRINSEC, *La Commutation Electronique,* Vols. 1, 2, Editions Eyrolles, Paris, 1980.

Chapter 2
Application Software for Telephone Switching Exchanges

INTRODUCTION

In the field of the software for telephone exchanges, the terminology used for the necessary concepts is not universally agreed. Therefore, this introduction gives most of the concepts in common use to reduce the risks of semantic misunderstanding. In so doing, a pragmatic approach has been taken, instead of trying to refer to global theoretical models, which would have made the considerations more precise, but at the cost of further obscuring topics that are already dry and hard to understand.

In the following sections, the application software of a generic exchange is described without referring to any specific system, nor to any conceptual model aimed at a universal description of the exchanges. The goal is to provide a common sense description of the realities that can be observed in the practice of switching exchanges. The details are retained only where they are essential to understanding the more general features. As the details may differ substantially from exchange to exchange, whenever they are referenced, they must be considered only as a meaningful example used for the sake of completeness.

The *exchange software* (see Figure 2.1), when analyzed from the viewpoint of the application transactions it executes, appears structured as an organic set of several *application activities* that are strongly interworking with each other.

Each application activity contains a set of functions having its own meaning and which can be described (in general terms) briefly and consistently without dealing with complex and painful details. The application activities mentioned here are software packages, the dimensions of which lay in the range of several tens of thousands of statements.

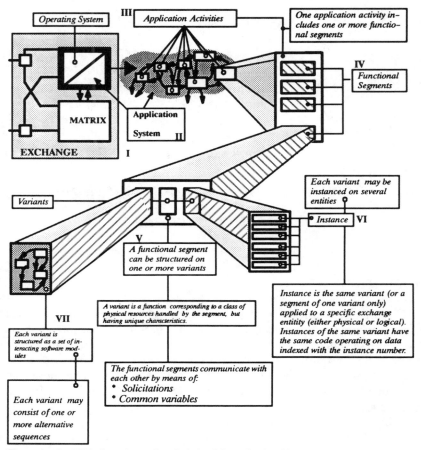

Figure 2.1 General structure of the exchange software application system.

An application activity can be structured into a set of one or more *functional segments,* each of which, as the word itself suggests, contributes to the economy of the entire activity by implementing a specific function. Each functional segment may include one or more *variants* describing the same function implemented in a different way because of the differences of the physical or logical entity of the exchange to which it applies. (For example, the functional segment dealing with the preprocessing of the incoming signaling may have one variant for each kind of line.)

Every variant of each functional segment may or may not be "*instanced*." An instanced variant may be applied at the same time to different physical or logical entities, which bear the same *characteristics* (for example, it can be applied to several subscriber lines with identical characteristics, to many bundles of the same kind, *et cetera*). An instanced variant is a package of software entities provided with a common index (either multidimensional or with one dimension only) that assumes a different value for each instance of that variant. A variant is "*not instanced*" when it allows one instance only.

As a particular case, an application activity may consist of one segment only; one segment can have one variant only; one variant can have one instance only or, equivalently, can be not instanced. In some circumstances, a useful approach is to organize each variant of a functional segment into a number of *sequences,* among which one can choose the one that must be executed each time. In these cases, if the variant is instanced, one can choose the sequence to follow, independently of every allowed instance.

The activities in which the application system is structured interwork and communicate with each other in a very intensive way. This is also true among the functional segments of the same activity. In our pragmatic model, the instance or instances of each functional segments are the active entities of the application system, which, loosely imagined for the sake of clarity, is similar to a set of people, each behaving according to the software described in a functional segment or variant of a segment. The instances of a multiple-instance segment are like people who operate in the same way.

These people, as well as the conceptual entities which describe their functional behavior, communicate with each other mostly by exchanging *solicitations*. A solicitation is an entity bearing the characteristic property of being sent from a *source entity* to a *destination entity* at a *given moment*. Having received it, the destination entity takes note of the fact and starts executing actions that depend on the received solicitation and parameter values that it may include. The same actions, however, may also depend on the values of specific variables, which may be read and updated by the same receiving entity. As a general rule, solicitations may be exchanged among any instance of instanced segments or by single-instance segments.

Each functional segment (or variant thereof), consists of a set of software structures defined according to system-dependent criteria. This level of detail is systematically ignored in our general approach here. However, each variant of single-instance segments and each instance of any instanced segment is seen as an *individual* capable of acting independently of the software that implements its behavior.

The set of activities part of the application system may be divided into two big families: *switching activities* and *operation and maintenance activities*. As the words suggest, the switching activities include everything related to the implementation of call processing; the operation and maintenance activities refer to the management of the exchange. (Note that in U.S. terminology "operation" often means switching.) In the following sections, the switching activities are described first, then the others.

2.1 Signaling Preprocessing and Postprocessing

A common control interfaces a multiplicity of different exchange terminations organized into peripheral units; each realizes the logical and functional interface between its terminations and the common control. Via the peripheral unit in which it is located, each termination exchanges its own *signaling* with the common control, characterized by semantic content, syntactic rules, and protocols, which vary from case to case. Together with the exchange terminations, the matrix and its marker (markers) are seen by the exchange software as additional peripheral units.

Note: In several system implementations, the marker rather appears to the common control as a set of several peripheral units. This point, however, does not affect the substance of considerations related to the operation of the matrix. Therefore, for the sake of simplicity, one can assume the existence of only one marker between common control and matrix. According to the actual situation, the marker will be either real or virtual.

The peripheral units may be organized into classes by placing in the same group all those which interact with the common control in the same way. For each class of peripheral units, there must be a specific activity in the application system that operates as an *intelligent driver* for every termination of that class capable of (see Figure 2.2):

- Preprocessing every incoming signal at its terminations to convert it according to syntactic requirements and expected protocols in the rest of the application system;
- Postprocessing every directive received from the remaining application exchange software to transform it into signaling to be accepted in the peripheral units.

The drivers have substantial processing capabilities because the protocols for the peripheral units differ substantially from those of the remaining application software, and this requires much more than a plain message transfer from one part to the other. Also, each driver must perform a set of functions necessary for the operation and maintenance of the units with which it interacts. In fact, as will be emphasized in the following sections, each driver must be capable of:

- Detecting, during the execution of its normal activities, the occurrence of anomalous syndromes that it will have to be able to transfer to the maintenance activities;

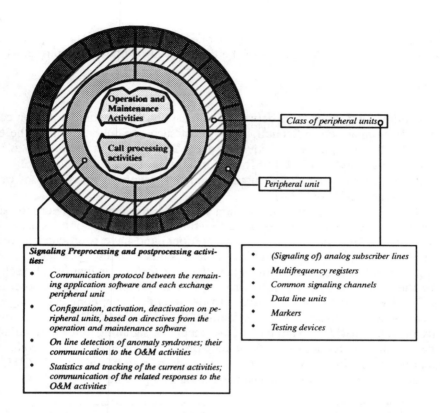

Figure 2.2 Signaling processing activities.

- Keeping track of its ongoing processing by means of suitable variables, to be able to communicate to the exchange operation activities the raw data to be used to measure what is actually happening in the exchange;
- Executing, upon indication from the operation and maintenance activities, all the relevant updates of the *configuration parameters* related to units and terminations. These parameters describe to the driver what to do and how to do it for each unit and termination;
- Activating and suspending, upon indication by the operation and maintenance activities, specific testing and tracking sequences for the signaling being exchanged with its terminations (depending on the case, on either some or all of them). The results of these actions will be sent to the relevant operation and maintenance activities.

In the following sections, each kind of driver is analyzed together with its features. Collectively, the drivers form (see Figure 2.2) a layer of interfaces between exchange terminations and matrix on one side, and the rest of the application software on the other. (A synopsis of the drivers is also shown in Figure 2.3.)

2.1.1 Subscriber Line Signaling

The communication procedures between analog subscriber line and common control have been analyzed in Section 1.3. Against such capabilities, within the exchange software, an application activity is needed, dealing with the *preprocessing of the telephone signaling* (PRETESI) and virtualizing the line units to make them appear as entities which employ a language compatible with the remaining software in the exchange (see Figure 2.4).

PRETESI is but one of a long sequence of acronyms introduced here to refer more easily to frequently used software functions. The introduction of these acronyms is based only on utilitarian considerations, and have been chosen freely from any reference to other terminology used in similar topics.

PRETESI basically deals with rectangular waveforms (see Figure 1.9, Section 1.3) of the incoming signaling leads from the terminations of analog subscriber lines. These waveforms are translated into the intended telephone signaling, or, as the case may be, into anomalous syndromes. With respect to PRETESI, another activity must be provided to implement the telephone signaling that the remaining application software requests to be forwarded on the outgoing leads of the exchange terminations of analog subscriber lines. This activity may be called ATECRI, as it provides for the *activation of telephone criteria* or signaling.

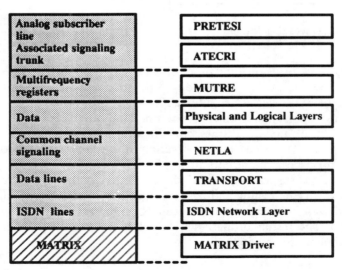

Figure 2.3 Terminal driver interfaces.

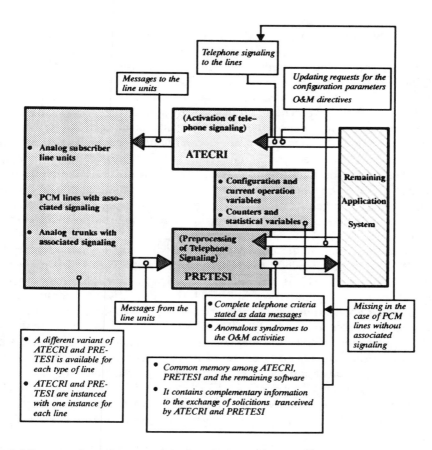

Figure 2.4 Preprocessing and postprocessing for telephone signaling leads.

Both PRETESI and ATECRI operate on a common set of variables that exhaustively describe the current operating conditions and configuration profiles of each analog line, its telephone terminals at the subscriber's premises, and its exchange terminations. Other variables are used with similar purposes for each unit in which the subscriber lines are organized. The set of these variables, which are typically referenced to the subscriber lines or the line units, constitute the reference model for the exchange software, describing each subscriber analog line and line unit.

The processing of PRETESI and ATECRI is based on both the solicitations that they receive from the remaining application software and the current values of the common variables considered here. The same variables may also be updated by PRETESI and ATECRI as a result of their processing. These activities, in fact, result in solicitations sent to the remaining application software and modifications of the variables that describe subscriber lines and line units. In every case, PRE-TESI communicates the telephone signaling to the remaining software by sending solicitations and updating common variables. In the same way, PRETESI and ATECRI specify the occurrence of anomalous syndromes and respond to the operation and maintenance functions in the exchange.

Some of the above-mentioned descriptors specify the *configuration profiles* of lines and units, and denote the services available to each subscriber. Other descriptors are used to state the *operating conditions* of the lines (e.g., in service, out of service, under test) and details related to their physical equipment (e.g., frame, rack, printed circuit board). All these descriptors are indirectly set and modified, via PRETESI and ATECRI, to which the remaining application software sends the relevant solicitations. Based on these solicitations, PRETESI and ATE-CRI consistently update the descriptors with the telephone traffic being carried on the lines.

Direct updating of the descriptors by activities other than PRETESI and ATECRI is typically avoided because this can cause problems of consistency for the processing of PRETESI and ATECRI, which also depends on the value of said descriptors. As already noted, PRETESI and ATECRI communicate with the remaining application software by exchanging solicitations and updating variables that are visible to both PRETESI and ATECRI and the remaining software.

Considerations similar to those for the case of analog subscriber lines apply as well for analog trunks and PCM lines with associated signaling. In these cases, in fact, the communication mechanisms between software and the lines are still substantially the same. Therefore, *PRETESI and ATECRI include all of the interfacing activities for subscriber loops and junctions with telephone lines, having their signaling structured as rectangular waveforms available on binary signaling leads.*

Because of the variety of such lines, PRETESI and ATECRI are structured into several variants; one for each type of line. *Each variant has a different instance for each line to which it can be applied.*

Similarly to the case of any other exchange device, PRETESI and ATECRI must also operate for each line a specific set of counters used in each circumstance and for that line to state the number of meaningful events for the operation and maintenance of the exchange. These counters are reset, started, suspended, and read to the remaining application software, based on solicitations that ATECRI and PRETESI receive. This is done to implement operation and maintenance functions.

In the case of PCM lines that do not use associated signaling, ATECRI and PRETESI deal only with the operation and maintenance of the units. However, the telephone signaling is handled by data multiregisters, switched in semipermanent way (nailed-up) on the relevant common signaling channel (see Figure 1.14 and Section 2.1.3).

2.1.2 Register Signaling

The communication procedures between multifrequency registers and common control have been analyzed in Section 1.2. To implement these features, the application system must include a specific activity for multifrequency signaling (MUTRE = *multifrequency treatment*). Due to the multiplicity of multifrequency signaling languages (CCITT R2 and national variations, CCITT 5, *et cetera*) and the differences between outgoing and incoming registers, MUTRE includes two variants (one for the incoming registers, the other for the outgoing ones) for each signaling language handled by the exchange. Each variant will have one instance for each register that it controls. A monitor must be placed above this instanced functional segment which manages the resources of MUTRE (see Figure 2.5).

Each time an application activity of the exchange requires a multifrequency register, the MUTRE monitor is activated, which then assigns one available instance in the relevant variant of MUTRE. Hence, the entity requiring the register, the seized instance, and its associated register are interlocked, and they remain so until the MUTRE monitor receives a release solicitation for the same instance and its associated register. The software entity that requires seizure of a register also states the parameters that it needs to carry the multifrequency activities.

Step by step, through interaction with its register, the instance of MUTRE carries out its activity by stating in a specific *ticket* the meaningful events that have occurred on the line. Also, in each relevant circumstance, the instance sends specific solicitations to the software entity that seized the register to read the data written in the ticket or to provide data needed by MUTRE to continue its activity.

Note: Normal events in a register are transmission and reception of information, such as seizure clear, digits from one to nine, extradecadic signaling (C11 and C12 *et cetera*). Abnormal events are basically unexpected downtime while transceiving signaling information.

When anomalous situations occur, MUTRE generates alarm solicitations to the operation and maintenance activities, which, upon reading the ticket, can decide what actions to take. As for the other exchange devices, each register is characterized by a set of:

- *Operating state variables,* which specify how the register is actually operating (out of service, under test, available, busy, *et cetera*);
- *Configuration parameters,* which state its functional, physical, and installation features;
- *Statistical variables,* which provide for a detailed indication of current working peculiarities. These variables are typically associated with threshold values, the exceeding of which causes the generation of alarm messages from MUTRE to the suitable operation and maintenance activities.

The modification of the operating states of a register is done by its associated MUTRE instance, which operates according to its interactions with the same reg-

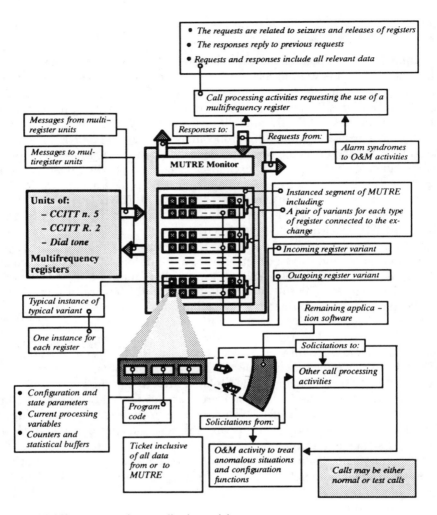

Figure 2.5 Multifrequency register application activity.

ister, following messages from the operation and maintenance activities. The configuration parameters are modified by MUTRE, following messages received from external application activities. The same is true as well for the reset, start, and transfer of the statistical variables.

2.1.3 Common Channel Signaling

The application activities related to the operation of common channel signaling are structured into three stratified functional segments. The first of these segments interacts directly with the channels, the *physical layer;* the second is the *frame layer;* the third is the *network layer.* The physical and frame layers have a different instance for each channel. The network layer consists of one instance only (see Figure 2.6).

The layered protocol permits the application processes in any switch to communicate with their counterparts (e.g., other switches, databases) through a set of transparent media. In the incoming direction from the channels, the physical layer deals with the synchronization of the incoming bit streams and verifies the actual operating conditions of the physical channels. In the opposite direction, the physical layer transmits uninterrupted, steady bit streams.

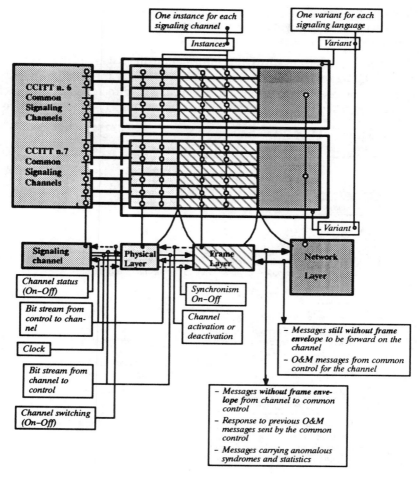

Figure 2.6 Interfaces for the activities related to common signaling channels.

In the incoming direction from the physical layer, the typical instance of the frame layer (see Figure 2.7):

- establishes frame synchronization;
- locates the incoming frames;
- verifies their correctness and discards faulty frames;
- forwards to the network layer the messages contained in the frames found to be correct.

In the opposite direction, the same instance generates to the physical layer a steady flow of frames, which may be:

- filled with messages received from the network layer;
- filled with messages generated by the same instance of the frame layer;
- devoid of useful messages.

The *network layer* receives the information messages in the frames of all instances of the frame layer (see Figure 2.8). For each information message received, the frame layer verifies whether the message is directed to its own exchange or must be forwarded to another switching exchange on a common signaling chan-

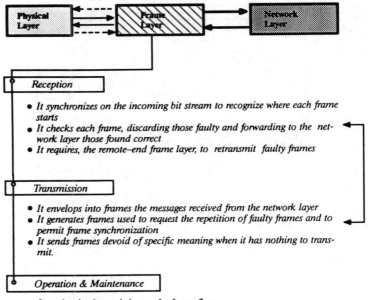

Reception
- It synchronizes on the incoming bit stream to recognize where each frame starts
- It checks each frame, discarding those faulty and forwarding to the network layer those found correct
- It requires, the remote–end frame layer, to retransmit faulty frames

Transmission
- It envelops into frames the messages received from the network layer
- It generates frames used to request the repetition of faulty frames and to permit frame synchronization
- It sends frames devoid of specific meaning when it has nothing to transmit.

Operation & Maintenance
- It maintains its statistics on the frame flow
- It informs the network layer about harmful anomalies
- It executes, upon directives from the network layer, sequences of channel start, restart, stop, and network management

Figure 2.7 Frame layer functions for a common signaling channel.

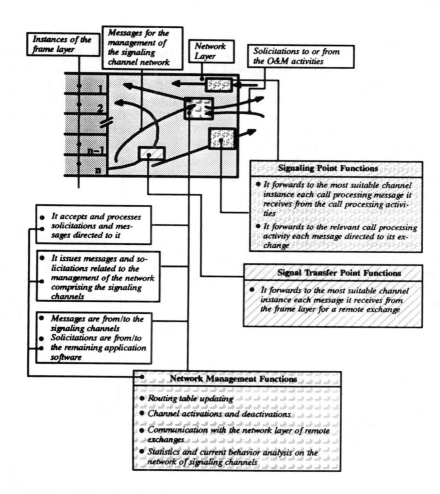

Figure 2.8 Network layer functions for common signaling channels.

nel. In the first case, the message is sent to the relevant application software activities. In the second case, the network layer modifies the address field in the message, if necessary, for properly forwarding it; then the network layer sends the message to the frame layer instance associated with the channel on which the message must be transmitted.

The network layer receives from other application activities of its exchange the messages to be forwarded to distant switching exchanges. The network layer selects the channel on which to send each message that is transferred to the associated instance of the frame layer. The frame layer adds flags and ancillary information to each message to make it become a frame. It is then passed to the same instance of the physical layer and from the physical layer to the channel.

The basic capability of the network layer is to decide which direction to forward each useful message that is received. For that purpose, the network layer uses specific routing tables, updated by itself, based on messages received from other application activities. To change the routing tables, the network layer employs protected procedures through which it changes the routing rules without disrupting the routing being executed.

The network layers on any pair of exchanges connected by one or more signaling channels communicate with each other according to normalized (i.e., standard) procedures. These procedures, consisting of interactive sequences of messages, allow the network layers of different exchanges to agree with each other about the routing rules that apply at any moment on the *signaling network,* comprising all common signaling channels connecting the switching exchanges. In any case, the *management of the common signaling channel network* is done by the network layer of the exchanges involved, and on the basis of rules stated in tables and parameters within each exchange. The tables may be updated in each exchange by the network layer, which operates on the basis of solicitations received from other application activities of the exchange operation software. The central point is that the purpose of the physical, frame, and network layers is to maintain an environment in which application processes (e.g., call routing) can communicate with other processes elsewhere in the network.

At the international level, two common channel signaling languages are standard (normalized): CCITT 6 and CCITT 7. For each language with which it deals, the exchange software will be provided with the relevant activities for the physical, frame, and network layers. The substantial differences between CCITT 6 and CCITT 7 imply corresponding differences in the related application activities.

Each common signaling channel communicates with the common control via digital multiregisters. (See Figure 2.9; see also Figures 1.7 and 1.14.) The signaling channels may be connected also through *modems* (see Figure 1.15). However, in the case of digital exchanges, this alternative is rather an exception, and therefore is not taken into consideration.

Digital multiregisters are implemented by using microprocessors and VLSI components, through which they include the function of both the physical and frame layers. Consequently, the application software actually loaded in the common control includes only the functions related to the network layer. This can be done by means of a specific application activity NETLA (network layer) that operates as the monitor for the network layer, includes one instance only, and manages all the multiregisters. NETLA may include two variants: one for CCITT 6, the other for CCITT 7. NETLA interacts with the physical drivers of the multiregisters, and also includes the operation and maintenance features associated with the multiregister units.

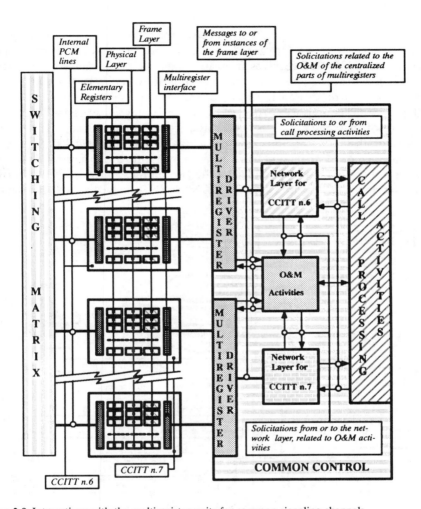

Figure 2.9 Interactions with the multiregister units for common signaling channels.

2.1.4 Data Links

A point-to-point data channel between an exchange and a remote processor is handled in the exchange by software activities that are structurally similar to the physical and frame layers of common signaling channels. In the case of these links, there is no network layer activity. Instead, there must be an end-to-end transport activity (TRANSPORT). This activity is the logical interface between the remaining exchange software and the remote processor, and uses the frame and physical layers of the data link, which provide protected and error-free communication on the transmission link.

TRANSPORT (see Figure 2.10) receives from the rest of the exchange software the messages to be forwarded to the remote processors. In the opposite direction, TRANSPORT receives from its frame layers messages for the rest of the exchange software. Through specific peculiarities that vary from line to line, TRANSPORT virtualizes each remote processor and makes it appear consistent with the procedural interfaces expected by the remaining software in the exchange. In the case of switched data links, such as packet-switched lines, TRANSPORT also includes the necessary mechanisms to make and accept calls. In this way, when a message needs to be sent from the exchange software to a remote processor, if there is not yet a connection between the two entities, TRANSPORT provides one. In the opposite direction, when TRANSPORT receives a call on a switched

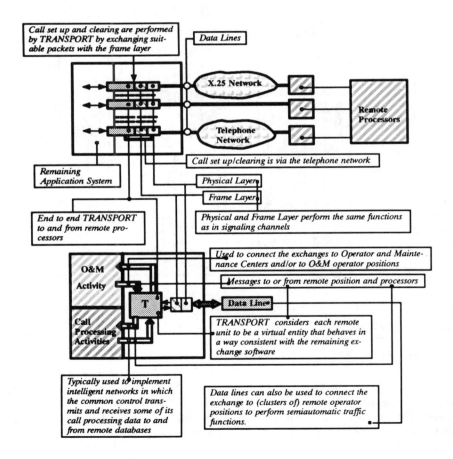

Figure 2.10 Modes of use related to data lines.

data line, it is accepted. Obviously, TRANSPORT is also able to end data calls and re-establish calls interrupted because of communication failures.

Note: Again, in this section, only the case of synchronous data lines is explicitly considered as this is the most natural choice in modern digital exchanges. The alternative of asynchronous links is similar and can be easily derived from the considerations here.

Switched data lines can be implemented by either circuit-switched telephone lines or X.25 packet networks. In the first case, TRANSPORT directly interfaces the line's physical layer, which emulates the dialing and answering functions that everyone implements on his or her telephone set. In the second alternative, TRANSPORT executes the call functions by exchanging suitable control packets with the frame layer. (This is done according to CCITT X.25 Recommendations).

Also, in the case of TRANSPORT, considerations related to the *configuration* and *operation* parameters apply. Configuration parameters are used by TRANSPORT as the information needed for correctly conducting its activities. Operation parameters give detailed tracking of what TRANSPORT is actually doing. Configuration parameters can be changed by TRANSPORT according to directives from the operation and maintenance activities. Operation parameters are made available by TRANSPORT to the operation activities for further processing.

Also, for the case of data lines, our concluding remarks about NETLA apply. Multiregister units are used to implement the physical and frame layers. TRANSPORT is an application activity of the common control; however, as will be shown in Chapter 3, TRANSPORT tends to be located remotely in the multiregister units.

TRANSPORT (see Figure 2.11) tends to have one variant for each kind of connection to remote processors. Each variant *V* is instanced with one instance for each data link implementing a connection of the same kind of *V*. For every data register, TRANSPORT has a physical driver in the central control where communication problems between the TRANSPORT functions and the remaining application system are handled. The driver also includes the operation and main-

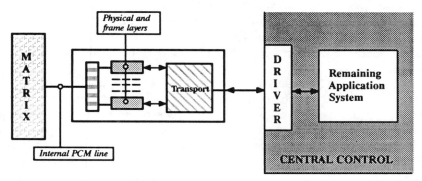

Figure 2.11 TRANSPORT implementation layout.

tenance function associated with the common parts of the multiregister unit. For that purpose, the driver exchanges solicitations and information with the operation and maintenance activities, similarly to the case of the other exchange termination units.

2.1.5 Signaling for ISDN Lines and X.75 Packet-Switching Lines

The D channel of an ISDN basic rate interface may be used both to transceive packet data and to signal control information for the two B channels of the same line. The structure of the data messages sent on a D channel is similar to that of common signaling channels, such as CCITT 6 and CCITT 7. Because of this situation, the exchange software for the basic rate interface must include an *ISDN physical layer* and an *ISDN frame layer,* similar to those for common signaling channels, even if different in major deails (see Figure 2.12).

In addition to the physical and frame layers, an *ISDN network layer* must be provided. This layer, in the incoming direction from subscriber to exchange, separates packet data from signaling. The network layer exchanges packet data with *packet-switching activities* (see Section 2.4) and signaling with the activities in the exchange software for circuit-switched calls. In the opposite direction, the ISDN network layer accepts messages from both packet mode and circuit mode switching activities. This information is sent to the frame layer for transfer to the D channel of the relevant line, via the physical layer.

In this case, considerations related to data, parameters, and counters for operation and maintenance purposes likewise apply as for the previous units. Also, ISDN basic rate interfaces are organized into line units comprising a number of microprocessors. These microprocessors (see Figure 2.13) naturally tend to include physical and frame layers. Therefore, within the exchange software loaded in a common control, only a unit driver is available, which concentrates and dispatches the messages related to the basic rate ISDN interfaces and includes the functions of the network layer. The same driver also includes the usual operation and maintenance functions. The layered architecture, as described earlier, should make all of this transparent to the network layer and, more importantly, to the rest of the application software.

Primary rate ISDN interfaces are PCM lines, where one channel is used as a 64 kbit/s D channel and the others (30 in the CCITT standard, 23 in the North American approach) are B channels. *The D channel can carry only signaling and not packet data.* Each primary rate interface terminates on an exchange unit, which is a PCM termination without line signaling (see the last paragraph of Section 1.6). The D channels of ISDN primary rate interfaces are switched in a semipermanent way (nailed-up; see Figure 2.13) on a specialized data multiregister akin in structure to those used for common channel signaling. These registers include the physical

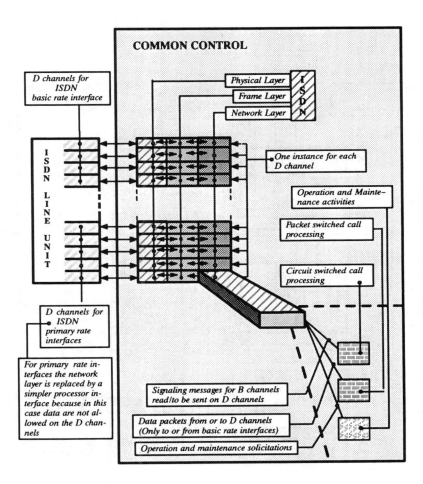

Figure 2.12 ISDN channel management.

and frame layers, but a *very reduced network layer* because, as already noted, D channels of primary ISDN carry only signaling information related to the B channels. For both basic and primary ISDN, B channels may carry packet data, and this is decided on a call-by-call basis. When this happens, a B channel is switched via the matrix on an X.25 or X.75 (see Figure 2.14) data multiregister structurally akin to those used in common channel signaling. This allows the common control proper to transceive data packets on the switched B channels.

Data communication can also be established between ISDN subscribers and subscribers of packet data networks, according to the CCITT X.25 Recommendations (see Section 1.11). To make this possible, an ISDN exchange is provided with X.75 junction lines used to forward calls from or to remote data packet nodes.

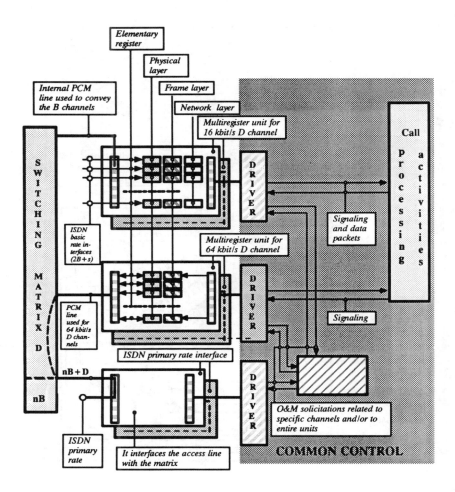

n = either 23 or 30

Figure 2.13 Application activities for ISDN access channels.

These lines are some of the 64 kbit/s channels conveyed on PCM lines (See Figure 2.14). They are switched in a semipermanent way (nailed-up) on X.25 or X.75 data multiregisters (identical to those used for the B channels of ISDN primary rate interfaces), which allow the common control proper to exchange packets with the lines, without dealing with problems related to the physical and logical layers. *Note:* The reader should again consider how the physical and logical layers adopted for CCITT 6, CCITT 7, ISDN D channels, ISDN B channels, and X.25 or X.75 differ and require specialized multiregisters, even if akin in nature.

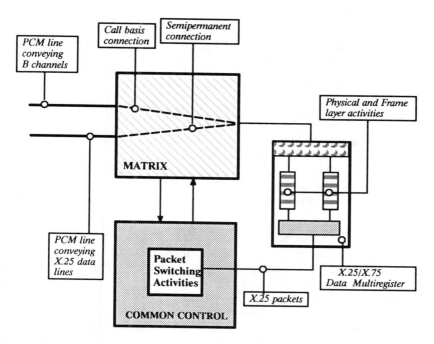

Figure 2.14 Application activities for ISDN access channels (B channel packet data).

The physical and the logical layers are software functions and, at least in principle, are part of the common control. However, as already noted, they tend to be implemented in specialized units. This is why they have been considered outside the *common control proper*.

2.1.6 Switching Matrix

Besides the firmware functions typically implemented within a switching matrix, the application system activities related to the operation of the matrix include the following points:

- Choice of the optimum path (if any) available at each moment to interconnect within the matrix its terminations, as required by the remaining application software. Subsequent execution of the circuit mode connections;
- Release of existing connections following specific requests from the remaining application software. The logical and physical resources used for the released connections become available for other calls;
- Detection from the firmware and hardware of the matrix of its functioning conditions. Forwarding to the matrix of *confinement* commands for the matrix

sections in which faults have been localized. Signaling to the relevant application activities of anomalous conditions found in the matrix;

- Execution, following solicitations from operation and maintenance activities, of test routines in the switching matrix. The results of these routines are sent to the activities that requested them;
- Introduction into service (upon solicitations from the operation and maintenance software by operator commands) of units of the matrix. The actual introduction into service occurs after the execution and positive results of *test routines.* If these test routines were to give negative results, the operator requests would be refused and a negative acknowledgement sent back to the application activities that had started the process;
- Execution of statistics to describe the actual traffic being carried by the matrix units.

The results of the statistics are made available to operation and maintenance activities. The switching matrix and its associated application activity in the common control interface with each other by means of a message protocol, which is different from system to system, but is structured around a set of elementary *commands* and *responses,* including specific, spontaneous messages from the matrix to indicate its actual operating conditions. As is typical in such situations, software and matrix interact with each other, given timing constraints, the overflows of which are meaningful indications of what is actually happening. The software activity related to the switching matrix interacts with the remaining application software to and from which it accepts or acknowledges connection setup and release requests. Also, the operation and maintenance functions include their own exchange of solicitations with software that controls the matrix.

As we will see in Chapter 3, all the application activities considered here naturally tend to be placed in a marker used as an intelligent buffer between matrix and common control. By doing so, the software of the common control proper includes only a driver that interfaces the marker with the remaining application software (see Figure 2.15).

2.1.7 Data Processing Peripherals

Ordinary data processing peripherals, such as printers, disks, and operator terminals connected to the common control, are handled by specific drivers, which, by analogy with telephone peripherals, may be considered as part of the application software. These drivers include the following functions:

- They implement a communication protocol between their peripherals and any application activity using them. (If more than one activity may seize the same peripheral at the same time, the driver also includes the ability to resolve concurrent access conflicts from more than one activity.);

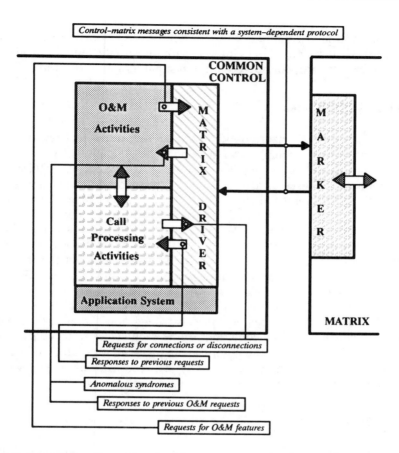

Figure 2.15 Interactions with matrix and marker.

- They execute, as in the case of any driver, an analysis of the current behavior of their peripherals by informing the related operation and maintenance activities of any possible anomaly;
- They carry out on their peripherals, upon directives from the operation and maintenance activities, routines of configuration, test, introduction in service and out of service.

The above-mentioned functions are unit-dependent and vary from system to system. No further comment is made on this topic as it does not show specific attributes that can be ascribed to switching exchanges.

2.1.8 Testing Devices

Section 1.12 has outlined the use of testing machines, which activate artificial calls to test both the exchange and its lines. Each of these devices finds a counterpart driver, similar in structure and performance to those already mentioned in Section 2.1.7. The drivers of testing devices interact with the specific *testing procedures* of the maintenance software, considered in the last paragraph of Section 2.5.5.

2.2 Circuit-Switched Automatic Calls

Each call, from the initial seizure on its incoming line to the final release of both called and calling line, is a complex transaction for the exchange software, resulting from the concurrent interaction of several software entities, each specialized in the implementation of a subset of actions. As well as successful calls, an exchange must deal with call attempts, which, for any reason, do not reach their expected destination. For each successful and unsuccessful call attempt, a *transaction* is defined in the exchange to which a *ticket* (i.e., call record) is associated. For the call attempt, its ticket identifies, with the necessary details, what must be stored and known about the call process. The ticket is generated when the transaction starts. The ticket's content changes during the call. The ticket is cancelled at the conclusion of the call attempt, and after the postprocessing of its data, to keep track of that call attempt as necessary. (Examples of postprocessing results for a ticket are the billing data to charge the subscribers and the statistical data to understand the actual operating conditions of the exchange.) A ticket is a *conceptual entity* having a strongly system-dependent physical structure. As a general rule, the ticket comprises one or several physical records, plus several variables distributed over dynamic structures such as queues and tables.

2.2.1 Main Processes

The first application activity, starting when a call is initiated (see Figure 2.16), is the *treatment of the signaling on the incoming line* (INLINE), where the call has originated. Conceptually, at least, INLINE is structured around a functional segment with a number of instances equal to the maximum number of call attempts that can be treated at a given time by the exchange.

Note: The maximum number of call attempts that can be handled at the same time by an exchange is a system parameter. This implies that each logical instance provided in application activities, such as INLINE, is a *logical resource* of the switching exchange. This resource can be used to its maximum availability in the exchange.

In addition to its instanced segment, INLINE has a further *dispatching segment* that receives from the remaining telephone software requests to start up new calls. The dispatching segment assigns to each request, depending on availability, an instance I of INLINE. The chosen instance number (i.e., I) is sent to the software entity which has made the request. These requests come to INLINE from the

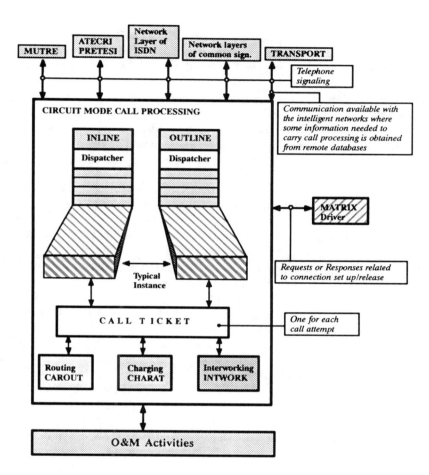

Figure 2.16 Circuit mode call processing activities.

signaling preprocessing activities; from PRETESI, NETLA, and the network layer of ISDN channels. In each case, the seizure request arrives at INLINE together with the coordinates of the incoming line where the call attempt is occurring. INLINE replies to such a request with either a negative acknowledgement (due to *congestion* situations) or with an instance I assigned by the dispatcher.

At the same time, the dispatcher of INLINE opens a ticket with the value of I and the coordinates of the incoming line written on it. This establishes a biunivocal link between incoming line and instance. Henceforth, instance I of INLINE can transceive complete telephone signaling on its associated incoming line, via the application activities that deal with preprocessing and postprocessing of the signaling. All the relevant events are then recorded on the ticket. By complete telephone signaling, we mean messages in which well defined telephone events are established such as *answer, release,* or *seizure.* These telephone events are described together with the parameters needed to define them completely. (An example of these parameters is the number of telephone subscribers.) Conversely, *messages which state only the occurrence of transitions on binary signaling leads are not complete messages.* The relevant preprocessing and postprocessing activities are PRETESI and ATECRI in the case of associated signaling lines, or the network layers in the case of common signaling and ISDN lines. The details of the activities implemented by INLINE are strongly dependent on the characteristics of the incoming line.

When such a line uses a multifrequency register, I of INLINE also invokes MUTRE, asking it for the seizure of a suitable incoming register. After having obtained such a register (if it does not, due to congestion, I of INLINE ends the call), I of INLINE writes its related instance Re on the ticket, then asks for the matrix driver to actuate the physical switching of the incoming line with the register Re. Afterward, I of INLINE sends Re of MUTRE a solicitation in which it asks for the execution of the relevant software sequence. At this point, Re of MUTRE begins to interact with I of INLINE, sending it all the relevant signaling data. When the register's signal processing is completed, I of INLINE asks for the disconnection of incoming line and register. This is so only in the case of incoming lines using multifrequency signaling. Otherwise, I of INLINE can directly receive the dialed digits and write them on its ticket. With or without register, at a given moment, I of INLINE arrives at the point where it has a number of dialing digits sufficient to try to route the call. When this happens, I of INLINE invokes another application activity dedicated to *call routing* (CAROUT), asking it whether routing is possible, based on the ticket content. If yes, CAROUT gives the address of the outgoing line on which the call must be forwarded. If CAROUT replies that the number of digits is not sufficient, I of INLINE will try again, after having received the next digit. The procedures according to which CAROUT computes its routings are considered in Section 2.2.2.

When it knows to which outgoing line to forward its call, I of INLINE calls for another application activity specifically for the treatment of the signaling on the *outgoing lines* (OUTLINE). INLINE and OUTLINE are conceptually similar to each other. Their differences are that INLINE treats each call attempt on the incoming line, while OUTLINE deals with the outgoing line of the call. For each instance of OUTLINE there is a corresponding instance of INLINE. This allows I of INLINE to ask OUTLINE to activate its corresponding instance, which, for the sake of simplicity, can be assumed as labeled by the same integer I. The I of OUTLINE finds all the information it needs on the ticket and can therefore start its actions on the outgoing line.

I of OUTLINE operates on its line via the relevant instances of the signaling preprocessing or postprocessing activities (i.e., PRESTESE and ATECRI, the network layer of the D channels, NETLA). Similarly to INLINE, OUTLINE also uses MUTRE if the line on which it is operating requires a multifrequency register. The goal of OUTLINE is to signal on the outgoing line the messages needed to forward the call. After activation, *I* of OUTLINE operates in close correlation with its parallel instance of INLINE. To do so, the respective *I* of OUTLINE and INLINE exchange solicitations that are interpreted by each, based on the data written on the call ticket, which is the same for both instances.

The processing by INLINE and OUTLINE strongly depends on the lines involved in the call. This means that both INLINE and OUTLINE have several variants; typically, there is one variant for each kind of line connected to the exchange. In this situation, the two lines involved in the same call and associated with the respective *I* of INLINE and OUTLINE may be of different types and signaling languages. In this case, the *I* of INLINE and OUTLINE are of different variants. However, in the application software, each variant of INLINE is designed on the assumption that it must interact with only the same variant of OUTLINE. In order to solve this contradition, another activity is available in the application software to provide *interworking* capability among different signaling languages. This activity may be referred to as INTWORK. Each time the respective *I* of INLINE and OUTLINE operate on different variants, they communicate with each other through INTWORK, which operates as interpreter or translator between them. By doing so, each variant of INLINE and OUTLINE operates in every case as if it interworks only with some sort of standard version of the corresponding activity. The functional structure of INTWORK is described in Section 2.2.3.

Each time INLINE and OUTLINE find the occurrence of events that prevent the successful completion of the call attempt, they release their lines according to procedures that depend on the line signaling language. However, when everything evolves properly and the called subscriber answers, a circuit switching is requested on the matrix. This is typically done following a specific request from OUTLINE, which is the first to realize that the called party has answered. After a call has been put through, the exchange may provide a charge for it. For that purpose, either INLINE or OUTLINE (the choice varies from system to system) starts another activity dealing with the identification of the *charging rate* (CHARAT) to be applied to the calls. CHARAT, based on the data written on the ticket, always gives the rate to charge. This information is used by OUTLINE or INLINE to update on the ticket the amount to be charged to the subscriber. A description of the criteria on which CHARAT operates is given in Section 2.2.4.

When either subscriber hangs up, or when the call ends for any reason, INLINE or OUTLINE sends to the matrix driver the request to interrupt the connection between the two parties. Afterward, the parallel instances of OUTLINE and INLINE carry out the relevant release procedures on their lines. In any situation, when the call ends, the ticket is sent, either by OUTLINE or INLINE, to the operation and maintenance activities, which will process the ticket for statistical and charging information.

2.2.2 Routing

A call is routed on the basis of an analysis of its dialing digits and other information such as the class of the incoming line used by the call. The outcome of the routing activities is always one of the following decisions (see Figure 2.17):

A The number of available digits is still *insufficient* to carry out the routing; more digits are needed.

B The digits received are *inconsistent* with the routing alternatives allowed in the exchange so that the call attempt must be cleared.

C The number received identifies a *subscriber* belonging to the exchange and provided with only one line; the call must be forwarded to that line.

D The number received identifies a subscriber with *multiple lines* connected to that exchange; these lines are typically connected to a PBX (private branch exchange).

E The number received identifies a *special service* offered by the operating company to its subscribers (such as telephone directory enquiry, customer premises equipment maintenance, *et cetera*) who dial special numbers to

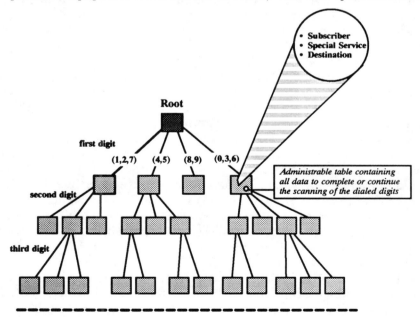

Figure 2.17 A routing tree example.

access them; CAROUT selects an outgoing line on which to forward the call; possibly subsequent digits may be handled by INLINE and OUTLINE.

F The call attempt must only *transit* through the exchange. CAROUT chooses the trunk on which to forward it. This is the most complex alternative. To treat it, CAROUT refers to *routing models,* which vary, at least in terms of details, from system to system. For the sake of clarity, a model of this kind is analyzed.

The universe of the places to which an exchange can forward the transit calls it receives is structured within the exchange software as a set of nonoverlapping *destinations* or *destination points* (see Figure 2.18). To make the point clear, a switch in Chicago might find the same list of routing choices appropriate for the following three dialed numbers:

1 212 777 7777	(New York City)
1 415 777 7777	(San Francisco)
011 441 777 7777	(London)

Although these are in very different locations, from the viewpoint of the local switch, they are all accessed through the same trunk group (bundle) because they all need to be routed to a toll center for recording and carrier selection.

The trunks of an exchange are organized into *groups* or *bundles* (see Figure 2.19). Each bundle is a set of trunks connecting the same pair of exchanges and seen by both of them as totally identical, especially with respect to the signaling language that they use. A bundle is either *bidirectional* or *monodirectional.* In the first case, its trunks can be seized by either termination exchange. In the second alternative, only the termination exchange which sees it as *outgoing* can seize its trunks. For the other termination exchange, the same bundle is seen as *incoming* in the sense that it cannot seize its trunks to forward transit calls.

A set of one or more bundles connecting the same pair of exchanges form a *route* when they are defined in the exchange software as equivalent alternatives to forward *every call directed to any destination.* Bundles of the same route may use different signaling languages.

A *direction* is a set of one or more routes departing from the same exchange, *E,* that *E* can use as alternatives to forward any call for any destination. The routes of a direction departing from *E* do not contain incoming bundles for *E*; they may arrive at different exchanges which *E* uses to forward its calls.

Destinations, bundles, routes, and directions are coded into tables within the exchange software. These tables are *administrable* in the sense that they can be modified (on line) by means of operator commands.

In order to perform its analysis on the called party's dialed numbers, CAROUT uses an administrable set of data tables, organized according to a tree-like structure (see Figure 2.17). At the root of this structure is a table that states the decisions CAROUT must make on the first digit of the dialed number. An arrow

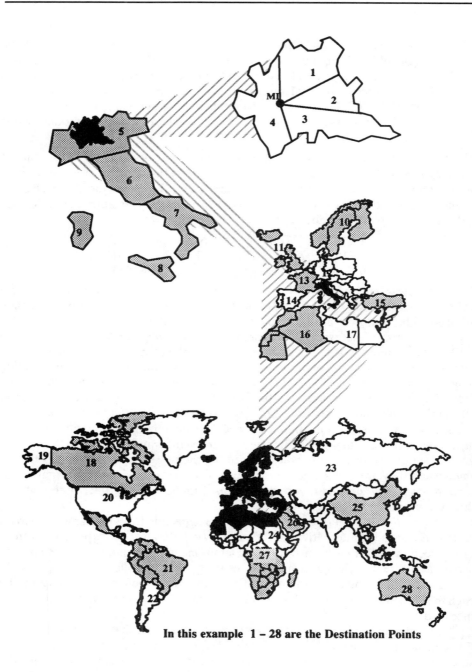

Figure 2.18 Concept of destination.

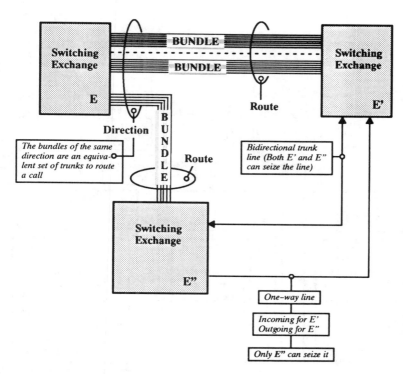

Figure 2.19 Routing entities.

from the root of each allowed value, if any, of the first digit leads to other tables that are structurally identical to the root. These tables are used for the analysis of the second digit. By recursion, the procedure is extended to the number of digits that a dialed number may include. Each table, at any level, allows for CAROUT to decide whether it must proceed to the analysis of a further digit. If not, it decides which option among those listed as A to F applies. If F applies, CAROUT also finds the destination point of the called party.

When the destination point is found, CAROUT must choose an available trunk line to forward the call. First (see Figure 2.20), a direction is chosen, then a route within that direction, a bundle, and a trunk. For each destination point, an administrable table in the software states the directions from which CAROUT can choose. These directions are ordered so that CAROUT can choose lower priority directions only after having checked that no trunk is available in higher priority directions. In each direction, its routes are also ordered by priority of choice as are its bundles within each route. Also, in these cases, CAROUT tries higher priority items first.

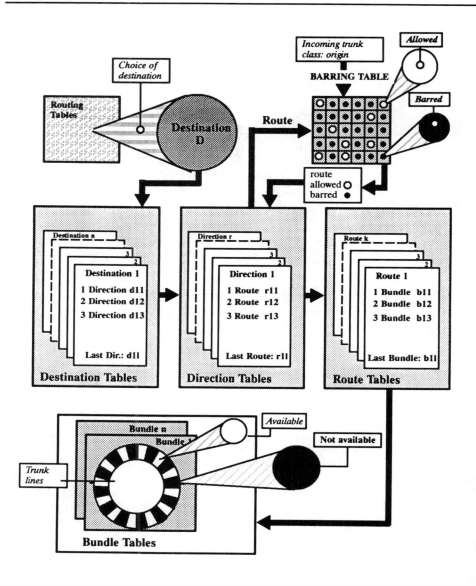

Based on the destination, an available line is chosen which belongs to the highest possible priority (hpp) bundle (as set in the Route Tables) of the hhp not barred route (as set in the Direction Tables) of the hhp direction (as set in the Destination Tables)

Figure 2.20 Routing choice of the trunk line.

In order to avoid the possiblity of forwarding the calls in transit into loops, the exchange software contains administrable *barring tables,* which, for each incoming bundle, states the routes where no call coming from that bundle can be forwarded in transit. These barring tables are analyzed by CAROUT for each call to check whether any route being considered is adequate. Switches also may use routing partitions, instead of barring tables, for outgoing and incoming traffic to avoid the looping problems mentioned here.

2.2.3 Interworking

INLINE and OUTLINE interwork on one side with the pre- or postprocessing of the trunk lines; the other side uses INTWORK. The interface with INTWORK has been defined in the CCITT Recommendations. On this subject, the adopted model is based on a set of *telephone events* divided into two groups: *forward interworking telephone events* (FITE) and *backward interworking telephone events* (BITE). (See Figure 2.21.) The FITEs are sent from INLINE to INTWORK and from INTWORK to OUTLINE. The BITEs go in the opposite direction. FITEs and BITEs have been defined identically for all the signaling languages expected from a telephone exchange. They are listed in Figure 2.22. Each variant of INLINE and OUTLINE associated with a signaling language generates and receives FITEs and BITEs according to on-line, real-time procedures specified for each language in the CCITT Recommendations. In an exchange provided with lines of languages $L_1, \ldots, L_n$, INTWORK includes $n \times n$ variants IW_{io} where IW_{io} is used by the variant i of INLINE to interwork with the variant o of OUTLINE. Each variant IW_{io} is instanced with a number of instances equal to the number of calls that may be active at the same time on the incoming lines of language line i with outgoing

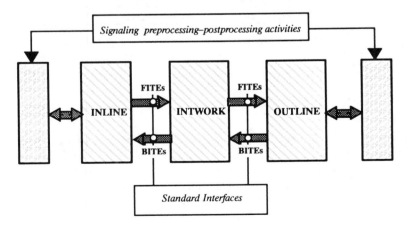

Figure 2.21 Interworking processes.

No.	Forward interworking telephone events
1	Digit 1,2 ...9 or 0, code 11 or 12, end–of–pulsing (ST) signal
2	Country–code indicator, country code not included
3	Country–code indicator, country code included
4	Echo–suppressor indicator, outgoing half–echo suppressor not included, incoming half–echo not required
5	Echo–suppressor indicator, outgoing half–echo suppressor included, incoming half–echo suppressor required
6	Country–code indicator, country code included; echo–suppressor indicator , outgoing half–echo suppressor not included, outgoing half–echo suppressor required
7	Country–code indicator, country code included; echo–suppressor indicator , outgoing half–echo suppressor not included, no echo suppressor required
8	Country–code indicator, country code included; echo–suppressor indicator , outgoing half–echo suppressor included, incoming half–echo suppressor required
9	Calling–party's–category indicator, operator, language French
10	Calling–party's–category indicator, operator, language English
11	Calling–party's–category indicator, operator, language German
12	Calling–party's–category indicator, operator, language Russian
13	Calling–party's–category indicator, operator, language Spanish
14	Calling–party's–category indicator, operator with forward–transfer facility
15	Calling–party's–category indicator, subscriber
16	Calling–party's–category indicator, subscriber or operator without forward–transfer facility
17	Calling–party's–category indicator, subscriber, ordinary call
18	Calling–party's–category indicator, subscriber, call with priority
19	Calling–party's–category indicator, data call
20	Nature–of–circuit indicator, no satellite circuit in the connection
21	Nature–of–circuit indicator, one satellite circuit in the connection
22	Clear–forward
23	Forward–transfer
24	Continuity

Figure 2.22(a) List of forward interworking telephone events (FITEs).

lines of language o. INTWORK, in addition to its instanced variants, includes a dispatching segment, which is invoked by the instances of INLINE each time any of them starts to interwork with an outgoing line. On the basis of the parameters given upon initiation, the dispatching segment of INTWORK selects an available instance IW_{io} and assigns it to the calling line. That instance remains seized until the call in transit is fully released (see Figure 2.23).

FITEs and BITEs have been defined to do the following:

No.	Backward interworking telephone events
1	Spare
2	Address–complete, charge
3	Address–complete, no charge
4	Address–complete, coin box
5	Address–complete, subscriber free, charge
6	Address–complete, subscriber free, no charge
7	Address–complete, subscriber free, coin box
8	Call unsuccessful
9	Call unsuccessful, switching–equipment congestion
10	Call unsuccessful, circuit–group congestion
11	Call unsuccessful, switching–equipment congestion or circuit–group congestion
12	Call unsuccessful, national–network congestion
13	Call unsuccessful, address–complete, national–network congestion
14	Call unsuccessful, address incomplete
15	Call unsuccessful, (address–complete), unallocated number
16	Call unsuccessful, address–complete, subscriber busy (elec.)
17	Call unsuccessful, address–complete, line out of service
18	Spare
19	Call unsuccessful, call–failure
20	Call unsuccessful, send special information tone
21	Answer, subscriber free
22	Answer, subscriber free, charge
23	Answer, subscriber free, no charge
24	Answer, reanswer
25	Clear–back
26	Artificial address complete may be sent
27	Sending–finished; set up speech condition
28	Deactivate register function

Figure 2.22(b) List of backward interworking telephone events (BITEs).

1. Minimize size and complexity of the IW_{io};
2. Make possible a univocal specification of the interfaces among INLINE, OUTLINE, and INTWORK;
3. Accommodate further extensions of the signaling languages.

The IW_{io} related to signaling languages made standard (normalized) by the CCITT are specified in the CCITT Recommendations.

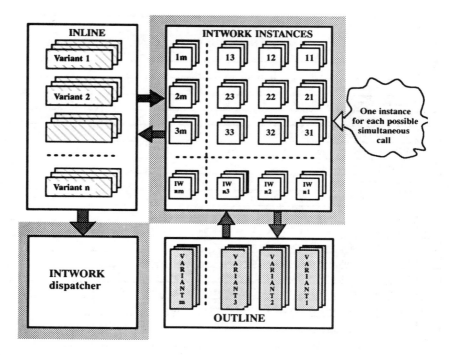

Figure 2.23 INTWORK structure.

2.2.4 Charging

Calls can be charged according to two different methods: by *time and charges* and by *pulse metering*. In the first case (North American standard), CHARAT writes the missing information on the call ticket to contain date, hour, minutes, and seconds of conversation start-up, conversation duration, and special service indicators. In this way, by postprocessing the call tickets at a later stage, to invoice the subscribers accordingly will be possible.

The case of pulse metering (European standard; also used in New York City and called message rate service) is more complex (see Figure 2.24). CHARAT, in the same way as CAROUT, obtains the destination point of the call. Then, by means of a suitable administrable table, CHARAT derives the *origin* of the call from the incoming bundle. By another administrable table, given origin and destination CHARAT obtains the *charging rate function* that applies to the call. Examples of charging rate functions are:

- No charge on this call;
- N pulses at conversation start, M pulses at call release, one pulse every S seconds during conversation.

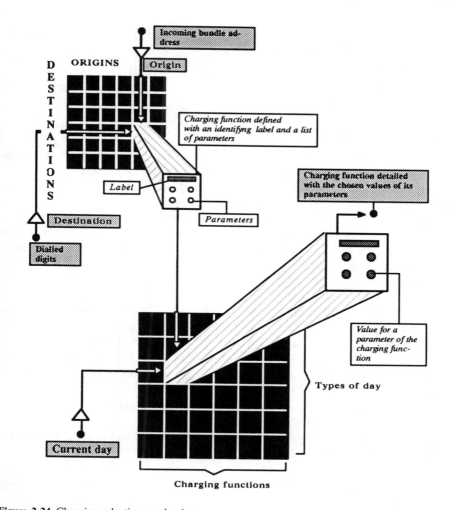

Figure 2.24 Charging selection mechanism.

The charging rate functions are labeled for use as indices to an administrable table, CHARGE, having as a second index the *type of day*. Types of day may be:

- Working day;
- Saturday;
- Sunday;
- Holiday.

Each element of CHARGE is a set of records, each stating for a given time span (from *xx.xx* hours to *yy.yy* hours) the values to be adopted for the parameters in each charging rate function (e.g., for the N, M, S of the previous example). The records included in each element of CHARGE fully cover, without overlapping, the 24 hours of a day. CHARAT, as a result of its processing, provides the relevant record of CHARGE and the charging rate function that applies to the call.

2.3 Semiautomatic Circuit-Switched Calls

In addition to automatic calls that are handled by the exchanges based only on actions usually carried out by calling and called subscribers, without any further human intervention, today some exchanges must still be able to treat *semiautomatic traffic* involving the active intervention of *telephone operators*.

Note: Manual or semiautomatic calls are treated differently from system to system. However, without giving some detail, the resulting analysis of this point would be too vague. For this reason, reference is made to a specific system used in the Italian public telecommunication network.

A semiautomatic or manual call, when analyzed in detail, appears as a sequence of *automatic elementary calls*. The first of these is a *booking call* from subscriber to operator in which a subscriber verbally tells an operator the kind of call he or she wants to place. As well as providing the indication of the called party, the caller may request *special services*, such as the following:

- *Person-to-person call.* A call to be directed to a specific person rather than a telephone number. Person-to-person calls are charged only when the called party replies.
- *Collect call.* These calls are charged to the called party after he or she has given the operator his or her agreement to pay.
- *Third-party call.* A call to be charged to a third party, someone other than the called and calling subscribers.
- *Credit card call.* A call to be charged to a credit card.
- *Appointment call.* A call to be placed at a given hour of the day. The calling subscriber may request that the call *not* be placed in certain periods of time.
- *Sequence of queueing calls.* The calling subscriber may ask for the operator to place a sequence of calls one after the other.
- *Time and charge information.* The calling subscriber requests to be called back by the operator at the end of the call to be informed about duration and charge of the call.
- *Waiting time.* The subscriber may ask to know the expected waiting time before the call can be made.
- *Time notification.* The calling subscriber asks to be notified by the operator when a given amount of time during the call is about to expire.
- *Completed call information.* The calling subscriber may ask to the operator for information about calls placed and completed during the prior few hours.

In all cases listed above, the operator must ask for and give information. After the first call from subscriber to operator, other calls will be needed from operator to subscriber, from operator to operator, and with or without requests (from operator to exchange) to connect or disconnect the subscribers involved.

2.3.1 Tickets and Files

Also, each semiautomatic call has its own ticket, which, however, is more complex than that of automatic calls. A semiautomatic call ticket, in fact, may include the following information:

- *Dates and durations* of the call, such as date and time of queueing, date and time of call start, conversation time, *et cetera*;
- *Subscriber indications,* such as indication of the trunk group used; subscriber number; name of the place where the call must be forwarded; routing chosen by the operator (i.e., manual line chosen, forwarding to C11 and C12 code operator (see Section 2.3.2), forwarding to a remote national operator, *et cetera*); routing chosen by the common control (i.e., trunk group used, direction of forwarding, *et cetera*);
- *Call characteristics* written by the operator (for a list of this point, see Figure 2.25);
- *List of the operator positions* that have dealt with the call;
- *List of the transactions* carried out by the operators on that call (such as call transfer from operator to operator, time and charge, *et cetera*);
- *Notes* written in plain language by the operators.

Number and extension of information written on semiautomatic tickets change from case to case and during the course of the call. This causes these tickets to be physical records of variable length. Semiautomatic tickets are placed during the course of the calls on disks, at specific locations of common control memory, or in memories associated with operator positions (see Section 2.3.4).

Like most administrative activities, telephone operators tend to organize their work into files. From an application software standpoint, a *transaction file* is an integer number within a range stated at the system level (for example, from 010 to 999). This set of numbers is a *logical resource* of the exchange. At any moment, a telephone operator may request from the system an available file number N. After having it, the operator will write that file number N on the ticket. From this moment, N becomes busy; and so it remains as long as the call associated with its ticket is in progress. The file numbers are used by both the operators of that exchange and those of remote exchanges to access directly semiautomatic calls in progress (which have been provided with a file number).

CODE	DEFINITION
X1	Station to station
X2	Station to person
X3	Collect to station
X4	Station credit card
X5	Station to person credit card
	$X = 0$: from subscribers
	$X = 1$: from public booth of class A
	$X = 2$: from public booth of class B
4.1	Marisat station to person
4.2	Ship station to person
4.3	Ship collect call
5.1	Service Call
5.2	Cancelled
5.4	Transit
5.5	Free call
5.6	Incoming call (C11, C12)
6.0	Incoming collect
7.0	Data transmission

Figure 2.25 Semiautomatic call facilities.

2.3.2 Incoming Call Queues

Each incoming call to an exchange operator, before being forwarded to an operator position, is temporarily waiting in a queue. There are several different kinds of queues. A possible list includes the following:

- One or more *booking queues*. Each time a subscriber dials for the semi-automatic service, he or she enters the exchange on an incoming line, and the application software puts such a call in a *booking queue*. There may be several booking queues, depending on the organization of the services offered to the subscribers. The software will choose the queue in which to put the call, based on the incoming line from which the call arises, some dialing digits received and configuration parameters internal to the exchange.

- Up to nine *C11 queues* (not available in the North American dialing plan). Operators connected to remote exchanges from all over the world may ask for assistance from the operators of the exchange involved. To do so, they dial a number including the extradecadic digit C11. (For the dispatching of semiautomatic traffic, besides the dial digits 0 to 9, two extradecadic digits *C11* and *C12* are available, which may be selected only at the operator positions.) By selecting a number that includes C11 followed by a number *h* (h = 1 to 9), the request is intended to reach an operator speaking language *h* (1 = English, 2 = French, *et cetera*). The receiving exchange, having found *C11 h* among the dialing digits, puts the call on the queue of operators that offer their services in language *h*.

- Up to nine *C12 queues* (not available in the North American dialing plan). Remote operators may request to the exchange being considered the intervention of operators specialized in offering a service *k* (k = 1 to 9) as agreed by the telephone companies. To do so, they include in their selection C12 followed by *0k*. The receiving exchange selectively places each call in up to nine queues specialized for the services *C12–01*, to *C12–09*.

- *Call assistance queue* for incoming international CCITT 5 lines. In a call forwarded on a CCITT 5 line, signaling requests can be placed for operator assistance. When these requests occur, their ticket flags are placed in this queue, which is like a special C12 queue.

- One or more *queues for national operators*. In addition to the queues for C11 and C12 typically used to locate calls from foreign operators, there may be similar queues for calls from national operators who do not use the C11 or C12 codes, but rather dial special selections that are different from country to country.

- Some *service queues* used to relocate calls already in waiting until a suitable operator can pick them up. Examples of these queues are:
  - *flash back queue* used to put calls in waiting between subscribers who are already in conversation when either party quickly presses and releases the telephone hook. Such an action is a way for a subscriber to ask an operator for help.
  - *notification queue* is used to put in waiting calls already in conversation for which an operator must inform the subscriber that an amount of time, previously established by the calling subscriber, is about to expire;
  - *time and charge queue* is used to locate completed calls for which the calling party has asked to be informed about charge and duration of the call.

- Several *manual line queues*. Especially in the case of international exchanges, manual trunk lines are still necessary on which, as a signaling criterion, only the seizure is available. These lines are typically organized into several bundles $F_1, F_2, \ldots, F_n$. To each F_i, a queue is associated on which the exchange puts the calls coming from that bundle.

The above-mentioned queues are FIFO (first in, first out) queues. The exchange selects from each queue the longest waiting call when an operator authorized to deal with it becomes available. In every case, putting further calls on each queue is suspended when either of the following circumstances arises:

- The number of calls on that queue exceeds a threshold set as a configuration parameter of the exchange;
- The longest waiting call on the queue has been doing so for a time exceeding a threshold set as a configuration parameter of the exchange.

An incoming call that cannot be put on a queue is refused according to normal procedures in the exchange to deal with congested conditions. Calls are put in the queues by an identifier, which states the trunk from which they are coming and on which a subscriber or operator is waiting.

2.3.3 Operator Services

Each exchange has a set of operator positions (POT) for semiautomatic traffic, which can be identified through a set of positive integers o_p ranging from 1 to the number of elements included in POT (see Figure 2.26). The working activities of telephone operators are structured into a set of disjoint operator services (SO), each defining competencies and responsibilities. Also, these services may be iden-

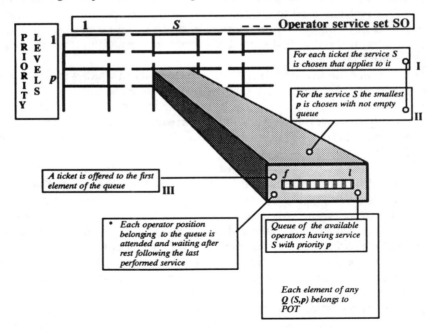

Figure 2.26 Operator queues for semiautomatic traffic.

tified by using integers ranging from 1 to the number of elements included in SO. To each operator position o_p an ordered list of services is assigned for which that position is authorized. Such a list can be identified as a string $S_{op} = (S_1, \ldots S_k)$, which specifies the service S_i given to o_p with priority i. Each S_i belongs to SO. Each o_p belongs to POT. The index k is a system constant (i.e., quantity stated once for all of the exchange software). Any pair of S_i belonging to the same S_{op} is made up of different elements. The S_i element of S_{op} is the service of priority i of o_p. If a service S belonging to SO is assigned to the operator position o_p as its service S_p (obviously $p \leq k$), S is assigned to o_p with priority p. The priority becomes higher as p becomes smaller. For each ordered pair (S, p), where S belongs to SO, and $p \leq k$, the set of all operator positions includes a subset $Q(S, p)$ made up of all operator positions o_p that have S with priority p.

At every moment, for each pair (S,p) in the application software, a queue is defined in which all operator positions are placed if both belong to $Q(S,p)$ and are *available*. Such a queue will have a first element $f(S,p)$ and the last element $l(S,p)$. The software distributes the tickets to the operator positions which are available at the moment, based on their current position in the queues of the sets $Q(S,p)$. More precisely, for each transaction that must be carried out on a semi-automatic call, the exchange common control will first decide which service of SO is needed. Afterward, the software will choose the *minimum p,* such that at least an operator position o_p of POT is available. An operator position o_p is considered available to execute service S with priority p, if and only if it belongs to the set $Q(S,p)$ and is located in the related queue of available positions. This fact is verified whenever position o_p:

- is properly operating and attended by an operator;
- is neither busy with another transaction, nor having a system administrable *rest pause* after finishing a previous transaction.

Each time more than one position is available, the one chosen is higher in the queue related to $Q(S,p)$. In each case, the chosen position o_p^* is removed from all queues, started, and put in a *not available condition,* where it remains until it has completed both its task and the subsequent rest pause (rest pause duration is defined as a system administrable parameter; i.e., as a parameter which can be modified by operation and maintenance activities). As soon as a position becomes available, it is put in the last place on the queues of the $Q(S,p)$ sets to which it belongs.

A typical list of a set of operator services follows:

00	No service,
01	National booking,
02	International booking with countries of group A,
03	International booking with countries of group B,

04	International booking with countries of group C,
05	Information,
06	Booking and forwarding for national calls,
07	Booking and forwarding for international calls for countries A,
08	Booking and forwarding for international calls for countries B,
09	Booking and forwarding for international calls for countries C,
10	Booking for calls from remote national operators,
11–19	Booking for calls from $C11h$ code operators (h = 1 to 9),
20	Forwarding only for national calls,
21–29	Booking for operator calls with code $C12h$ (h = 1 to 9)
30	Forwarding only for international calls,
31–60	Booking and forwarding on a specific group of manual lines,
61	Notification to the subscriber that the time set for their call duration is about to elapse,
62	Time and charge communication to the subscribers,
63–92	Forwarding of calls placed on special deferred call queues,
93	Data communication,
94	Forwarding of calls with a (transaction) file number,
95	General service (typically used as overflow emergency service).

2.3.4 Ticket Files

Each ticket, after being generated during acceptance of a queueing call, either stays *attached* to an operator position or is *placed in waiting* in some *ticket file*. At any given moment, an operator position can have attached to it only the call ticket on which the operator is working at that moment. Ticket files may be conveniently classified into the following types:

- Operator position files,
- Deferred call files,
- Appointment files,
- Special deferred call files,
- Files for calls outgoing on manual lines,
- Completed call files.

Hereafter, meanings and characteristics of each class of files are analyzed.

2.3.4.1 Operator Position Files

Two ticket files are associated with each operator position that the operator can manipulate by the keyboard. They are tickets *in hold* and *blocked*. A ticket associated with a semiautomatic call for which no conversation is in progress on the exchange lines may be protected by any operator by maintaining it in hold in a specific file associated to his or her position. An operator can manipulate the tickets in hold, sorting them on the terminal screen, one after the other. For a ticket already on the screen, an operator can forward the deferred call stated on it. This means that he or she can ask the common control to select the two parties, according to the subscriber numbers written on the ticket, over two outgoing trunk lines from the exchange. When subscribers or operators are reached, the operator involved can ask the common control to connect them or to start a three-party conversation with both of them. After a call has been established, the operator may *release* the call, which, from that moment is handled by the exchange without further operator intervention. Alternatively, with either one or both parties already connected, the operator can block the call. When this is done, the related ticket is placed on an operator queue, from which the same operator may ask for it later. For each blocked call, the operator position receives indications from the common control about any telephone event occurring on it. In every case, an operator can sort and read, one after the other, its tickets that are either in hold or blocked.

2.3.4.2   Deferred Call Files

In the treatment of semiautomatic calls, there are typical situations in which, for any reason, the operator cannot forward a call. In these cases, as already noted, the call ticket may remain in a file belonging to the same operator. This choice, however, is more of an exception than the rule. In general, in fact, deferred calls are put in centralized files by every operator position. Typically, there are four kinds of files for calls to be deferred to a later time:

- Urgent international calls,
- Urgent national calls,
- Normal international calls,
- Normal national calls.

The calls put in queue are later forwarded by the common control to operator positions.

2.3.4.3   Appointment Call Files

Subscribers may be offered the possibility to make appointment calls to be placed at a certain time in the following 24 hours. For this purpose, two FIFO files may be defined, for example, for each 5-minute interval of the 24-hour period; one for national calls, the other for international calls. The relevant tickets are placed in these queues to be forwarded to the operators when their time arrives.

2.3.4.4 Special Deferred Call Files

Among all possible destinations that may be reached by an exchange dealing with semiautomatic traffic, a useful option is to have a system administrable set of *special destinations. Destination is a set of country codes which can be reached by the exchange in the same way, and following the same strategy, as if they were the same place.* To each special destination, one or more deferred queues can be associated where to put the relevant tickets. These tickets will be placed in their deferred queues instead of those described in Section 2.3.4.2. Together with each special queue, a specific operator service is defined for forwarding of calls to each special destination. This gives the possibility of specialized service of special destinations from the entire accessible world. (The services for special deferred calls are numbered 63–92 in Section 2.3.3.)

2.3.4.5 Files for Outgoing Calls on Manual Lines

For each group of manual lines, in addition to a specific operator service (i.e., those numbered 31–60 in Section 2.3.3), a ticket file is available in which to put the ticket of the calls that will be forwarded on this group of lines.

2.3.4.6 Completed Call Files

Calls already completed are put in this file for possible further inquiries about completed calls by the operators, following specific requests from the subscribers.

2.3.5 Application Activities for Semiautomatic Traffic

Each semiautomatic call is started by a *booking call,* which may be generated by either a remote operator or a subscriber. This call is handled by INLINE, similarly to any other incoming call. In this case, however, INLINE, based on the selection digits received and parameters describing the trunk line where the call has arrived, identifies the incoming queue in which to put it. To unload the ticket queues, the application system must include a specific activity, which may be called TICKUN (TICKUN = ticket unloading) (see Figure 2.27). This activity sequentially scans the incoming call queues and ticket queues to find which one is not empty and therefore needs to be serviced. In each queue that is not empty, TICKUN sorts the first element and chooses the operator service necessary to deal with it. Based on such a service, TICKUN chooses the operator position that must deal with the ticket.

To implement the operator position functions, the application system must include an activity for the *treatment of the semiautomatic tickets* (TRESETI), having

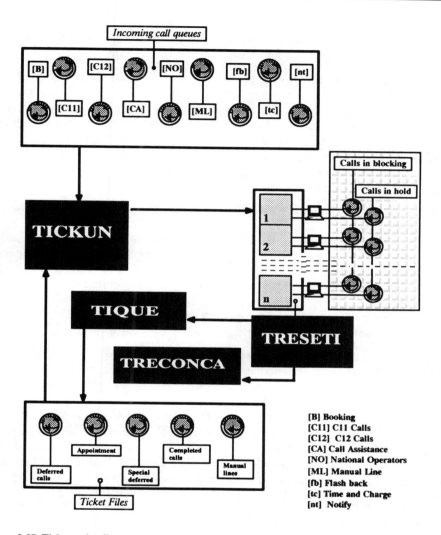

Figure 2.27 Ticket unloading system.

one instance for each operator position. Obviously, to send an incoming call or a waiting ticket to an operator, TICKUN activates the relevant TRESETI instance and transfers the call or ticket to it. In order to implement the relevant transaction, TRESETI does the following operations:

- It shows all information necessary at the operator position for the operator to carry on his or her job;
- It requires the matrix driver to connect the trunk line to the operator position cord;

- It signals to the pre- or postprocessor of the trunk line on which the call is waiting that the line has been connected to an operator.

When the above-mentioned operations have been executed, operator and calling party can talk to each other. At the same time, TRESETI generates the semiautomatic ticket of the new call. Based on the information typed by the operator, it fills out the same ticket. This completes the *booking* phase of the call. At this point, the operator can either try call forwarding or ask the calling party to hang on. If so, it sends the ticket on one of the waiting ticket queues. In every case, TRESETI operates based on the basis of what it finds written on the ticket and the commands received from the operator keyboard. Each TRESETI instance:

- Commands the matrix driver to carry out the relevant connections and disconnections;
- Interacts with the signaling pre- or postprocessors of the trunk lines, asking them to forward the telephone signaling and allowing its telephone operator to call subscribers or remote operators;
- Modifies and updates the semiautomatic ticket associated with its operator position;
- Requests another *ticket queuing application activity* (TIQUE) to transfer in the relevant file the ticket associated with its operator position.

TRESETI also carries on special functions such as ticket searching when an operator must comply with a request from a subscriber who asks to know about an already completed call. In this case, TRESETI, by interacting with TIQUE, finds the ticket and shows it to the operator.

TIQUE is not instanced and receives the queuing requests from all the instances of TRESETI. As already stated, TICKUN, TRESETI, and TIQUE may exchange with each other the same ticket several times, as it may need more successive phases to be completed. At the end, a TRESETI instance will close the semiautomatic call. This can be done in two ways:

- unsuccessfully, when called and calling party cannot be put in conversation;
- successfully, when they can converse.

In the second case, the operator releases the call to another application activity specialized in the treatment of the calls put in "cut through." This activity may be called TRECONCA (*treatment of connected call*) and operates similarly to INLINE and OUTLINE, but with the following peculiarities:

- it deals with both lines of the call;
- it writes the charging data on the semiautomatic ticket;
- in some cases (for example, when a *notification* or a *time and charge* has been requested by the subscriber), at the right moment it puts the call in an incoming queue;

- if the ticket is *blocked* at a given position o_p, it notes for o_p every telephone event of the call and sends call and ticket back to the instance o_p of TRESETI, upon request from the relevant operator;
- when a call finishes, it passes the ticket to the ticket file for completed calls.

Even the above-mentioned application activities, as well as performing their *useful telephone functions,* keep track of any anomalous situation by counting with suitable variables the events which are meaningful for both the operation of the exchange and that of the semiautomatic service.

2.4 Packet Switching

The scenario within which packet-switching functions must be included in an ISDN switching exchange have been discussed in Section 1.11. To understand the application software that is needed to perform these functions, we need to refer to the features and facilities actually performed in packet switching. This is to be done with some level of detail. Therefore, in the following section, a concise description of the packet-switching function is provided by referring to the CCITT X.25 Recommendation, which is specifically included in the ISDN scenario. Afterward, the structure of the application activities dealing with packet switching follows naturally, and is discussed in a subsequent section. To be concise, only the essential features of X.25 are discussed. Further details can be found in the relevant CCITT Recommendations.

2.4.1 Packet-Switching Functions

On the physical link between a subscriber *data terminal equipment* (DTE) and an exchange operating as *data communication equipment* (DCE), several calls can run concurrently. The same holds true on any link connecting two DCE so that on the same link (see Figure 2.28) more calls may be operational at the same time, typically between different pairs of subscribers. This is possible because, in each packet, a variable is present, called a *logical channel* and ranging from 1 to 4095. Different calls going through the same physical link are assigned a different logical channel to be written in their packets. A point-to-point call between two DTEs going through one or more DCEs is assigned on each link a logical channel, which can be chosen on each link independently of the others.

There are two types of packet calls (temporary) *virtual calls* (VC) and *permanent virtual circuits* (PVC). A virtual call is set up and released at DTE request; a permanent virtual circuit stays on indefinitely and is assigned to a pair of subscribers on a contractual basis. In VCs, the necessary logical channels are assigned at call setup and are released when the call ends. In PVCs the logical channels are assigned in a semipermanent way (nailed-up).

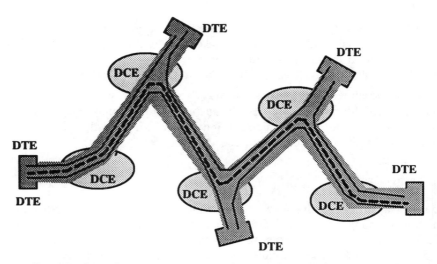

Example of use of logical channels to perform multiple connections between DTEs

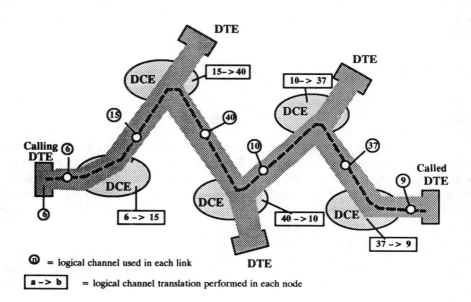

Figure 2.28 Use of logical channels in X.25 networks.

Packets may be divided into two classes: control and data packets (see Figure 2.29). Control packets are used to convey all signaling information needed to establish, control, and release the calls. Data packets convey the data exchanged between called and calling DTEs. In some cases, some control packets also may include user data. Each packet contains a *header* followed by a *variable* part.

Bit 1 of the header's third byte is 1 in control packets and 0 in data packets. The first half of the header's initial byte is a *general format identifier* (GFI) structured as shown in Figure 2.29. The second half of byte 1 is used to specify a group of logical channels. Byte 2 identifies a logical channel within a group. These two fields together form a 12-bit *logical channel identifier* (LCI) ranging from 1 to 4095. Packets that do not use a LCI (i.e., the restart, diagnostic, and registration packets) have their LCI set at 0. The third byte in the header is the *packet type identifier* (PTI), coded as shown in Figure 2.30.

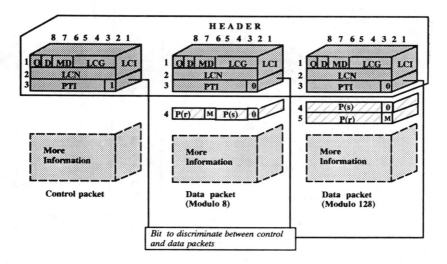

- The first three bytes of each packets are its **HEADER**
- Bits 8,7,6 and 5 of byte 1 form the **GFI** *(General format identifier)*
- **LCG** *(Logical channel group)* concatenated with **LCN** *(Logical channel number)* form **LCI** *(Logical channel identifier)*
- **PTI** *(Packet type identifier)* states the nature of the packet and the format of subsequent information
- **D** delivery bit – **Q** qualifier bit – **M** more data bit
- **P(s)** = *Packet send sequence number* **P**(r) = *Packet receive sequence number*
- **MD** = *Sequence numbering scheme* *(01 = Modulo 8; 10 = Modulo 128)*

Figure 2.29 General packet format.

Packet type		PTI bits							
DCE → DTE	DTE → DCE	8	7	6	5	4	3	2	1
Call set–up and clearing									
Incoming call (INC)	Call request (CAR)	0	0	0	0	1	0	1	1
Call connected (CAC)	Call accepted (CAC)	0	0	0	0	1	1	1	1
Clear indication	Clear request	0	0	0	1	0	0	1	1
DCE clear confirmation	DTE clear confirmation	0	0	0	1	0	1	1	1
Data and interrupt									
DCE data	DTE data								
DCE interrupt (INT)	DTE interrupt (INT)	★	★	★	★	★	★	★	0
DCE interr. conf.(INTC)	DTE interr. conf.(INTC)	0	0	1	0	0	0	1	1
		0	0	1	0	0	1	1	1
Flow control and reset									
DCE RR (Modulo 8)	DTE RR(Modulo 8)	★	★	★	0	0	0	0	1
DCE RR (Modulo 128)	DTE RR (Modulo 128	0	0	0	0	0	0	0	1
DCE RNR (Modulo 8)	DTE RNR(Modulo 8)	★	★	★	0	0	1	0	1
DCE RNR (Modulo 128)	DTE RNR (Modulo 128	0	0	0	0	0	1	0	1
	DTE REJ (Modulo 8)	★	★	★	0	1	0	0	1
	DTE REJ (Modulo 128	0	0	0	0	1	0	0	1
Reset indication	Reset request	0	0	0	1	0	0	1	1
DCE reset confirmation	DTE reset confirmation	0	0	0	1	1	1	1	1
Restart									
Restart indication	Restart request	1	1	1	1	1	0	1	1
DCE restart confirmation	DTE restart confirmation	1	1	1	1	1	1	1	1
Diagnostic									
Diagnostic		1	1	1	1	0	0	0	1
Registration									
	Registration request	1	1	1	1	0	0	1	1
Registration confirmation		1	1	1	1	0	1	1	1

Note – A bit which is indicated as "★" may be set to either 0 or 1

Figure 2.30 General packet types and PTI coding.

The *call request packet* (CAR) is issued by a DTE demanding the setup of a VC. This packet (see Figure 2.31) contains called and calling DTE numbers as well as the logical circuit chosen by the calling DTE on its link to the exchange. Upon receiving a CAR, a DCE decides the physical link on which it needs to forward the VC. On this link, the DCE selects an available logical channel that it replaces on the received CAR before sending it on the chosen link. This process is accomplished in all the DCEs from calling to called DTE. The DCE connected to the called DTE renames the CAR as an *incoming call packet* (INC) before

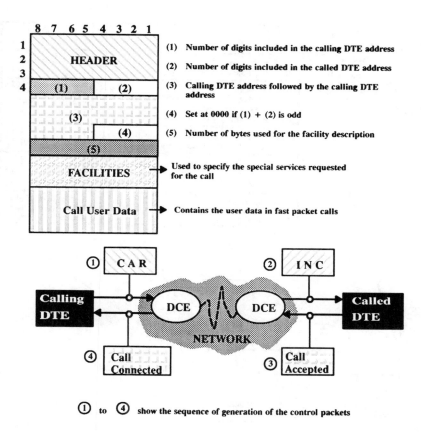

Figure 2.31 Call setup packets.

forwarding it. The called DTE, upon receiving an INC, replies with a *call accepted packet* (CAC), using the same logical circuit contained in the received INC. This CAC may also contain called and calling DTE address. The CAC packet moves back to the calling DTE, following, in the opposite direction, the same path previously established by the CAR. The CAC packet uses in each link the same logical channel of the corresponding CAR. At the end, the CAC packet arrives as a confirmation at the calling DTE, and the VC is established.

In the call set up process, wherever a relevant packet cannot be forwarded as needed, a CLEAR indication packet is generated. This packet follows the part of the path already established in order to clear it. The CLEAR packet contains a code stating the cause (see Figure 2.32) and contains a logical channel indication that is treated within each DCE in the same way as for the call setup packets. The same philosophy applies when any DTE or DCE, with a call already in the data

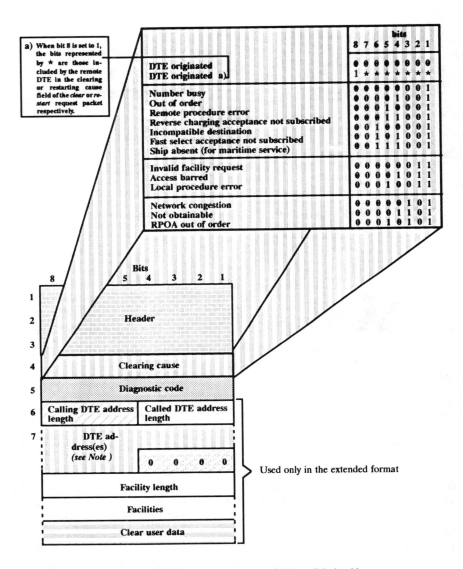

Note – The figure is drawn assuming that the total number of address digits is odd.

Figure 2.32 Clear packets.

transmission phase, finds some circumstance preventing a normal evolution of the call.

When either party wants to close a VC, it regularly uses a CLEAR request packet. Also, in this case, the CLEAR request contains the logical circuit referring to the call to be completed. The DCE replies directly to the closing DTE, confirming the request with another CLEAR packet. At the same time, that DCE sends a CLEAR indication to the next node involved in the call. The closing DTE behaves in the same way as the closing DTE and expects the same CLEAR confirmation. In this way, the clearing process propagates from DCE to DCE or DTE, until the other DTE involved in the call is reached. At this point, the call is completely cleared.

As soon as the call setup procedure has been successfully completed, data packets may flow in both directions. Every data packet (see Figure 2.28) bears a logical channel number, which is:

1. Set initially by the sending DTE;
2. Read by any receiving exchange to understand the call to which it is related;
3. Converted in each DCE from the incoming link value to the corresponding outgoing link value;
4. Read by the receiving DTE to understand the call to which it is related.

Both for PVCs and VCs, the packets are cyclically numbered by the originating DTE in the field $P(s)$ (see Figure 2.28). Numbering is accomplished by modulo 8 (i.e., the successor of 7 is 0). It may also be done by modulo 128 for users who have subscribed to this facility. The chosen modulus is coded in the GFI field of each packet. The numbering of packets is accomplished independently on each PVC or VC and is meant as a device to perform a flow-control procedure. This procedure may be accomplished either on a link-by-link or end-to-end basis by the two DTEs involved in each call. When the packets contain a D bit set at 0 (see Figure 2.29) only a link-by-link flow control is accomplished. When $D = 1$, the end-to-end flow control occurs. Hereafter, the case in which D = 0 is considered first.

For each direction of every link and every logical channel active on it, a *circular window* is set (see Figure 2.33) with two boundaries L and $L + W$. The *size*, *W*, of each window is constant within each call. At any moment, for each direction of transmission (e.g., the eastbound direction), L is the value contained in the field $P(r)$ of the last packet received for the same call, in the opposite direction (i.e., westbound). Based on the actual value of L, in the eastbound direction, to send only data packets with their field $P(s)$ equal to L, $L + 1$, . . ., $L + W - 1$ is permitted. Normally, while these packets are sent eastbound, other packets are flowing in the opposite direction bearing new values of $P(r)$, which, loaded in L, make the window move clockwise. If any congestion arises, either no westbound packet is received, or these packets contain constant values of $P(r)$.

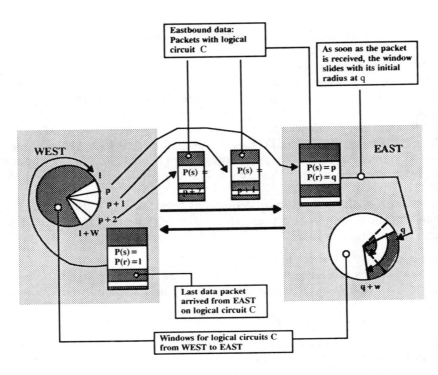

Figure 2.33 Window mechanism adopted in X.25.

This fixes L so that, as soon as a packet with $P(s) = L + W - 1$ is sent eastbound, no other can be transmitted.

A westbound packet contains a $P(r) = p$ after a packet has been received (for the same logical circuit) at the eastbound side of the link, which bears a $P(s)$ field equal to $p - 1$. The $P(r)$ field is included in all data packets so that those going in one direction may send flow-control information for the other direction. Also, special packets are available specifically for flow control. They are (see Figure 2.34) the *receive ready* (RR) and the *receive not ready* (RNR) packets. RR with $P(r) = p$ is sent westbound so that the eastbound window can be moved with its L equal to p. RNR is sent westbound to suspend (momentarily) the eastbound packet flow. RNR must hence be followed by an RR packet for the eastbound flow to resume.

In packets with the D bit set at 0, the flow control mechanism is carried out individually at each link in the path connecting the two DTEs. When $D = 1$, the users are requesting an end-to-end confirmation. Therefore, the values of $P(r)$ and $P(s)$ are generated by the DTEs, which act as if there were only one link directly connecting them. The *window size, W,* is a parameter that may vary from call to call, according to procedures analyzed in the following paragraphs.

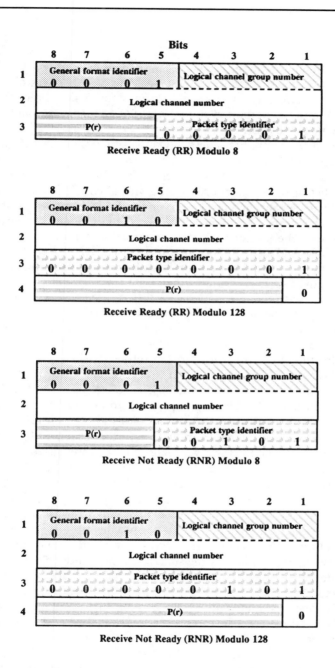

Figure 2.34 RR and RNR packets.

On a call already set up, either DTE can send an *interrupt packet* (INT) (see Figure 2.35) that bypasses the flow control mechanism and arrives directly at the other party. This is a feature used to send urgent data, which, however, must be small in size. A reply is permitted by an *interrupt confirmation packet* (INTC), which may contain some short data.

During the flow of data between two DTEs, situations may arise in which to suspend and resume the exchange of packets is necessary, without breaking the call. This is done by using the RESET packets (See Figures 2.36 and 2.37). The DCE or DTE that needs to start a reset on a virtual circuit (either a permanent virtual circuit or one established for a VC) sends a RESET request to its adjacent DCE or DTE. This request contains a code for the failure and the virtual circuit to which it applies. The packet navigates the virtual circuit until it again reaches the DTE/DCE that had started the whole procedure. The effect is that of again setting $L = 0$ of all the windows on the virtual circuit path. After the reset has been completed, the two DTEs resume their data transmission.

Packet retransmission is an optional facility offered to the subscribers on a contractual basis, and is common to all the logical channels of a DTE. Such a DTE may request the retransmission of data packets from its DCE by sending a reject packet (see Figure 2.38) specifying logical channel number and a $P(r)$. The value of $P(r)$ states for the DCE where to start the retransmission.

Each individual DTE is assigned on a contractual basis a set of available logical channels, which is a subset of the 4095 possible on a physical link. These channels are organized into disjoint groups, each to be used for either PVCs or VCs. Each PVC channel is defined together with the other DTE to which it is permanently connected. VC channel groups are further specified for (1) outgoing only, (2) incoming only, and (3) two-way calls. (In any case, there is bidirectional data flow between called and calling DTE. However, outgoing calls, originated by the DTE involved may not use its incoming-only logical channels. Similarly, a DCE, to forward a call to a DTE, cannot use its outgoing-only logical channels.)

Each DTE is assigned a *default throughput class* (see Figure 2.39), which applies to its available logical channels and specifies the bit rate expected on them. Each DTE is contractually granted a *load factor,* which states the maximum value that can be attained, at any time, by the sum of the throughput classes of its logical channels actually busy with calls. Any successive call that will cause an overflow in the load factor of a DTE is refused and cleared by the exchanges. During call setup, the DTE may negotiate with its DCE the actual throughput class desired for that call. This is done according to procedures discussed in the following paragraphs. In the negotiation, the throughput classes are individually defined (and may, therefore, be different) for each direction of transmission. In any case, the actual throughput class assigned to each call depends on the characteristics of both called and calling DTE, and is established during call setup. In making reference

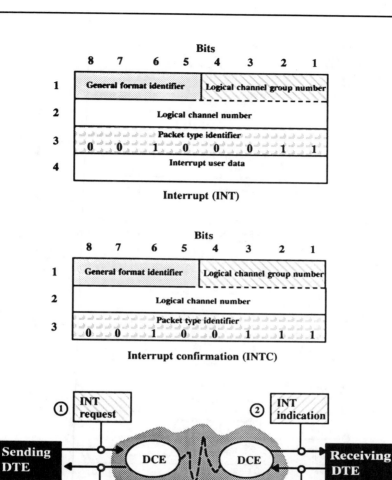

Figure 2.35 Interrupt packets.

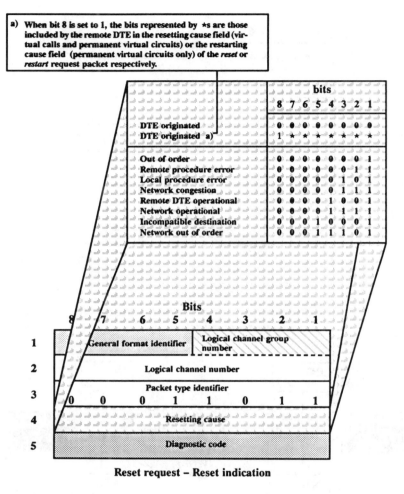

Figure 2.36 Reset packets.

to the load factor of a DTE, an exchange sums not the default throughput classes, but their final values, which result from the negotiation process.

The maximum number of data bytes that may be put in a packet is a system parameter, the standard value of which is 128. Other values are also possible, and they are 64, 256, 512, 1024, 2048, and 4096 bytes. The actual maximum packet length may be negotiated at call setup, as will be discussed in the following paragraphs. In any case, a DTE can put in its packets any amount of data up to the maximum value. When larger data messages are split into several packets, even smaller than allowed, the *more data bit M* is used. This bit is set at 1 in all the

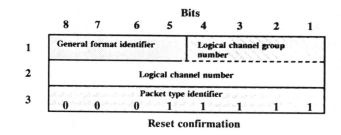

Reset confirmation

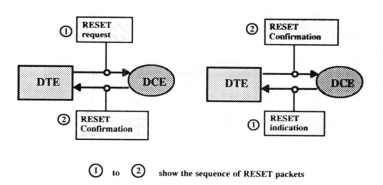

① to ②   show the sequence of RESET packets

Figure 2.37 Reset packets.

packets but the last one of the message, where it is set at 0. The exchange must be capable of treating these *sequences of packets* in the sense that they ought to merge consecutive packets smaller than the maximum allowed.

Note: Another special bit placed in the data packets (see Figure 2.29) is the *qualifier bit Q*. This bit is used by a DTE to divide data packets into two groups: one to be considered as ordinary, the other as bearing *special meaning*. The exchanges, as long as they treat only X.25 data, are transparent to the Q bit.

As a general rule, a call is structured into two phases: first, a virtual circuit is established; then, packets are made to flow on it. To save time in the case of very short calls, *fast select calls* can be implemented between DTEs that have subscribed to this optional facility. In fast select calls (see Figure 2.31), the CAR packet contains a fast-switching facility and up to 128 bytes of data. Similarly, the CAC packet from the called DTE also contains up to 128 bytes of data. In this way, the packets used for call setup also bring the data. Actually, in this case, they do not establish a virtual circuit as the CAC, alternating from called to calling DTE, closes the call.

Situations may arise in which an entire DTE must be restarted. Similarly, a DTE may require its DCE to make a global restart on its link. To do so, RESTART

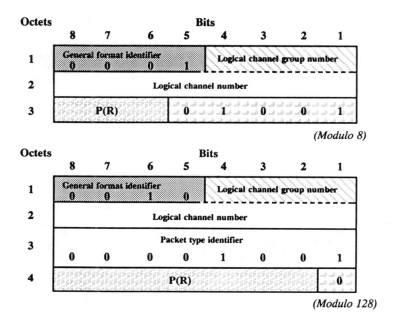

Figure 2.38 Reject packets.

Throughput class	Bit rate
0	75
1	75
2	75
3	75
4	150
5	300
6	600
7	1200
8	2400
9	4800
10	9600
11	19200
12	48000
13	48000
14	48000
15	48000

Figure 2.39 Throughput class definitions.

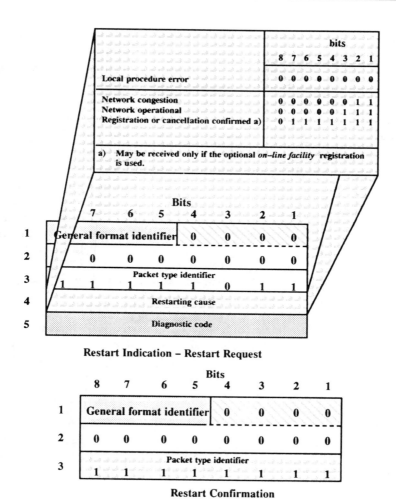

		bits							
		8	7	6	5	4	3	2	1
Local procedure error		0	0	0	0	0	0	0	0
Network congestion		0	0	0	0	0	0	1	1
Network operational		0	0	0	0	0	1	1	1
Registration or cancellation confirmed a)		0	1	1	1	1	1	1	1

a) May be received only if the optional *on–line facility* registration is used.

Restart Indication – Restart Request

Restart Confirmation

Figure 2.40 Restart packets.

packets are used (Figure 2.40). A restart procedure causes the release of all VCs in progress on that link and a reset of all PVCs.

A *facility field* is present in packets such as call request, incoming call, call accepted, call connected, clear request, and clear confirmation. This field is used to state special services that are requested with a call. Some of these services (or *facilities*) have already been mentioned. They are:

- *Window size negotiation* by means of which a DTE sets for the window a value W other than the one otherwise assumed as default value by the exchange;

- *Throughput call negotiation* by means of which a DTE selects a nonstandard throughput class, individually for each direction of transmission;
- *Maximum packet length negotiation* to select an admitted length other than the standard 128 data bytes;
- *Fast select call.*

The configurations used for the facility field to specify the above services, as well as the others to be described, are not analyzed here for the sake of brevity. They can be found in the CCITT X.25 Recommendation.

Other facilities which may be requested in the facility field are:

- *Reverse charge and reverse charge acceptance* when the calling party wants the call to be charged to the called party;
- *Network user identification* is a password that the DCE requires from the DTE to accept an outgoing call. This is only for DTEs who have subscribed to this facility;
- *Charging information* is to be sent on the CLEAR confirmation packet;
- *Selection of the transit network* via which the call should be routed.

DTEs may belong to *closed user groups* (CUG) with limited access capabilities other than among their members. A DTE may belong to one or more CUGs. The CUG to consider when making a call is also coded in the facility field.

All the facilities mentioned above, as well as a few others that have not been mentioned, apply only to users that have subscribed to them. Which DTEs are allowed to make use of which facility is part of the information stored in the configuration parameters of the exchange.

An X.25 DTE may register on line the facilities to which to subscribe on a contractual base. This must be done according to procedures using *registration packets* (see Figure 2.41). Other packets, such as the *diagnostic packets* (see Figure 2.42), are used in the X.25 procedures. Also, these points are only briefly mentioned here for the sake of brevity.

2.4.2 Packet-Switching Application Software

The packet-switching activity interacts, first (see Figure 2.43), with the drivers of the data multiregisters used for B channels, and X.75 trunk links. Similarly, packet switching interacts with the drivers of the ISDN basic rate interface. Through these drivers, the unit transceives the control and data packets considered in the previous section. From the same drivers, the unit also receives other messages about if and how any link fails. The packet-switching activity also interacts with the matrix driver to ask, by means of suitable messages, for any B channel to be switched on or off via a data multiregister whenever a packet call is about to begin, or when it has just been completed. In this case, the usual interactions with the operation

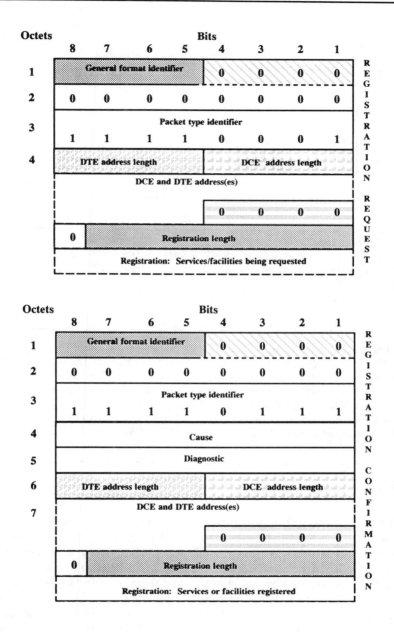

Figure 2.41 Registration packets.

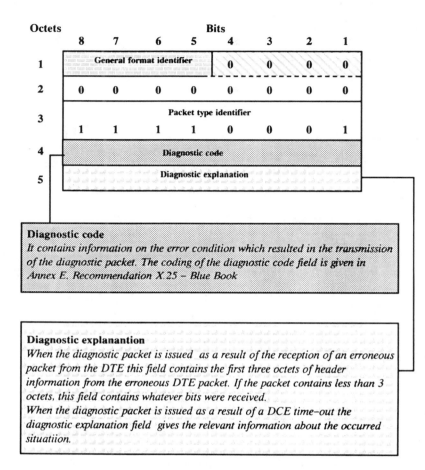

Figure 2.42 Diagnostic packets.

and maintenance software apply. Some sort of common data table maps which ISDN channels and X.75 links are connected to which registers for the packet-switching activity. In this way, it is possible to associate every incoming packet to its proper link. The association between registers and links, when done in a sem-ipermanent way, is handled by the operation and maintenance software; otherwise, it is done by the circuit-mode call processing functions.

The packet-switching activity operates on a finite set of *virtual circuits,* each of which is used to accommodate either a VC or PVC going through the exchange. These virtual circuits are a logical resource of the exchange. Their number is the maximum number of VCs and PVCs that may be active on the exchange at the same time. Each virtual circuit has a ticket associated with it that is empty when

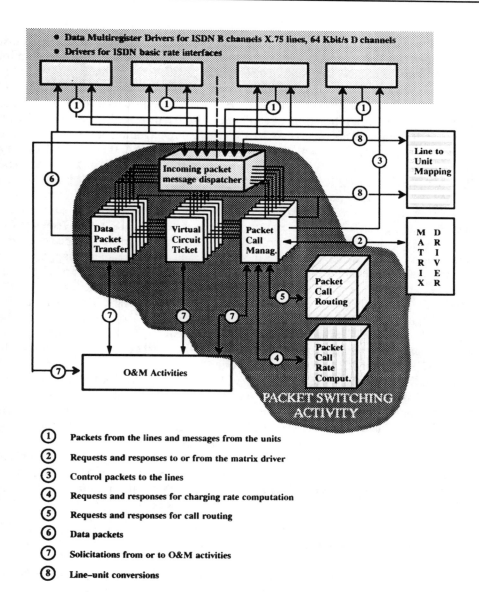

- Data Multiregister Drivers for ISDN B channels X.75 lines, 64 Kbit/s D channels
- Drivers for ISDN basic rate interfaces

Incoming packet message dispatcher

Line to Unit Mapping

Data Packet Transfer

Virtual Circuit Ticket

Packet Call Manag.

Packet Call Routing

Packet Call Rate Comput.

O&M Activities

PACKET SWITCHING ACTIVITY

M A T R I X D R I V E R

1. Packets from the lines and messages from the units
2. Requests and responses to or from the matrix driver
3. Control packets to the lines
4. Requests and responses for charging rate computation
5. Requests and responses for call routing
6. Data packets
7. Solicitations from or to O&M activities
8. Line–unit conversions

Figure 2.43 Packet-switching application activity.

the circuit is idle. In any other case, the ticket contains, among many other items, a logical circuit and link identifier of the two sides connected by its virtual circuit.

A first functional segment of the packet-switching activity is an *incoming packet dispatcher* (IMPADIS), which actually receives every packet and message arriving from the link drivers and the operation and maintenance software. Under IMPADIS, two other functional segments must be provided: one dealing with packet call setup and clearing (PACMA = packet management); the other for the implementation of the actual transfer of data packets on their virtual circuits (DAPACTRA). DAPACTRA also implements the data-flow-control functions. PACMA and DAPACTRA are instanced, with one instance for each virtual circuit. IMPADIS is not instanced. IMPADIS forwards every message and every packet it receives to the relevant instance of PACMA or DAPACTRA. In some cases, such as when a CAR (call request) packet arrives, IMPADIS actually selects the instance (i.e., the virtual circuit) to which to send the packet. In other cases, the same message or packet is sent to more than one instance. (This happens with restart packets and any other item that, affecting more than one virtual circuit, must be sent to all the concerned instances.) Basically, PACMA receives all packets and messages dealing with call setup and call clearing. DAPACTRA handles the others. Under PACMA, two other functional segments (not instanced) must be provided: one dealing with *packet call routing* (PACROUT), the other with *packet call charging rate computation* (PACRATE). Similarly to the case of circuit switching, PACROUT and PACRATE are invoked by PACMA each time during a call setup that their services are needed. PACMA and DAPACTRA send their messages and packets outside the packet-switching activity, as they need to do so to perform their functions. The segments store their ongoing data on the tickets related to the virtual circuits, which, similarly to the case of circuit-switched calls, are also used by the operation and maintenance software.

The treatment of PVCs is akin to that of VCs. The difference between the two cases is the fact that, for PVCs, the messages to the relevant PACMA instances mostly come from RESTART packets and the operation and maintenance activities.

2.5 Operation and Maintenance

Operation and maintenance features are implemented by the exchange software through a multiplicity of activities, which can be classified in the following groups:

S *Supervision and management* of the common control integrity;
O *Operation* of the remaining parts of the exchange;
M *Maintenance* of the remaining parts of the exchange;
I *Person-Machine interactions.*

The features under point *S*, at least in principle, are part of the *operating system,* which, in the case of telephone exchanges, must virtualize the common control to make it appear to the application system as a stable, real entity with no faults and no modifications. Point *S* includes the supervision of the current operation of common control, the implementation of the necessary reconfigurations, the isolation and putting out of service of faulty elements, and the introduction and activation of new and repaired elements. All these functions are strongly system-dependent and they are considered in Chapter 4, after the analysis of the architectures adopted for common controls.

The functions listed under points *O, M,* and *I* belong to the application system, and are therefore analyzed in Sections 2.5.1 to 2.5.3. Moreover, Sections 2.5.4 to 2.5.6 describe the application activities performing these functions in the exchange software.

2.5.1   Operation Functions

The operation functions cover the following broad classes of features:

- Configuration and reconfiguration of the parameters used in the exchange, by means of which the entire application system comes to be, as well as organization and characterization of every logical and physical resource in the exchange;
- Gathering of the charging data, traffic data, and statistics related to the traffic flowing through the exchange.

The above classes of features are executed by the software concurrently with those related to traffic management. In the considerations developed here, by *physical resource,* we mean any exchange device. Conversely, *logical resource* is any entity (such as trunk bundles and routing directions) that is meaningful for exchange operation, but not directly associated to a specific circuit entity.

As already mentioned, the exchange software knows the physical and functional realities of its exchange by means of *configuration parameters,* which are variables specifically defined for that purpose. Obviously, these parameters must be kept continuously consistent, in the sense that none of them can contradict its current values, those specified by the concurrent values of the others. As a general rule, a parameter can be either monodimensional (one variable only) or multidimensional. In every case, its value cannot be changed without taking the necessary precautions. In fact, indiscriminantly changing a parameter could jeopardize software activities, which, after having checked that a parameter had a given value, would see it change in ways incompatible with what they were doing. The configuration parameters are also said to be *administrable,* in the sense that they can be changed by means of procedures set by entities (i.e., either persons or machines) external to the exchange.

Parameter modifications may be originated by:

- Local operator requests;
- Remote operator requests;
- Reconfiguration activities that are part of the maintenance software (see Section 2.5.2).

In every case, each modification request causes specific procedures to start, which first check for the consistency and feasibility of the demand, and, in the negative case, give a suitable negative acknowledgement to the entity that had generated the demand. Afterward, only those requests which are found consistent are carried out by the same procedures, without interruption of the exchange operation. As a general rule, the second part of the above-mentioned procedures interacts with the activities that use the parameters which must be modified.

The updating of a parameter used by only one application activity is typically executed by the same activity. The operation procedure, in fact, will book a *parameter updating sequence,* which will be executed after the others have been momentarily suspended, or when they are not active. The updating of parameters used by other activities is much more complex. This circumstance minimizes the changes of parameters used by more than one activity of the application system. For each of them, the updating procedures are specifically and carefully designed. In general, when a parameter that affects several application activities must be changed, these activities receive a specific warning that causes them to go into a standby condition. When they are in standby, the parameter is updated. Then the concerned activities resume their work.

The measurement of the traffic flowing into the exchange is carried out by introducing, *enriched sequences* in the software activities related to signaling and call processing, which, while carrying on the traffic, sum the events to be measured in counter variables. These enriched sequences typically consume more machine time than ordinary ones and are started, changed, and stopped by directives issued by the application software activities dealing with *measure implementation.* These activities also request readouts and resetting of the counter variables, according to administrable frequencies. The software activities operate based on requests originated at operator positions and deliver the results of the measurements to files (to be subsequently accessed) and operator positions. In this manner, the operation software may execute the requested measurements by delivering the results to operator positions, printers, and mass memories. In Section 5.7, a type of internationally normalized procedure is introduced for the implementation of traffic measurements.

Also, *ordinary sequences* of signaling and call processing add significant events on (a reduced set of) counters used to determine current operating modes of the exchanges. *Similar counters are also used within the operating systems to evaluate the current loads of processors.* These counters are treated by operation software functions, similarly to those related to the implementation of measurement procedures.

The gathering of charging data, at least from a conceptual viewpoint, can be seen as a particular kind of measurement. It is executed by a specific software activity, which processes the call tickets by providing billing (raw) data in a mass memory. Based on these raw data, and by using processors external to the exchange, the telephone operating companies produce invoices for subscribers or other administrations. In the case where the charging data are established for each subscriber by means of *pulse metering or message rate service,* the exchange uses subscriber counters in which the activities dealing with call processings can add the total pulses for each call. In this case, the operation software will only transfer, at dates stated via operator positions, the current values of the subscriber counters in a mass memory. This is done on files, which are then used to bill the subscribers.

2.5.2 Maintenance Functions

Following errors on the lines or failures in the exchange, the application software must be capable of:

- detecting the occurrence of faulty situations (*failure supervision*);
- executing, following failure syndromes, analytical *verification procedures* on the exchange operational condition;
- locating and isolating, as precisely as possible, the consequences of faults (fault isolation or *first-level diagnosis*);
- providing tools to facilitate a precise and quick removal of faulty elements (repair verification procedures or *second-level diagnosis*).

This is done without interruption or substantial degradation in the exchange, with the possible exception of highly unlikely events, as long as the exchange is properly maintained.

In order to permit higher global availability levels, any faulty exchange device which causes serious misfunctioning is provided with some kind of redundancy. Consequently, when first-level diagnosis locates an error in a redundant unit, the procedure also provides for the automatic reconfiguration of the exchange by putting out of service the device found to be faulty and switching over its traffic to a standby unit.

Anomaly supervision is done, depending on the case, following three different approaches, which can be defined as:

- *continuous supervision,*
- *call-basis supervision,*
- *time-basis supervision.*

Continuous supervision is basically done by the hardware exchange and its firmware: any device that is critical for the overall operation of the exchange includes specific circuits that continuously probe its actual functional condition. The results of these probes are processed by firmware devices, which, in this way, continuously analyze the processing conditions of the device. Whenever an anomalous condition arises, the firmware generates signals perceived by the common control as *failure syndromes*. These syndromes are handled by the maintenance software. Continuous supervision is very costly because it tends to make the exchange hardware heavier, and is therefore limited to those functions which cannot otherwise be tested with the necessary timeliness for anomalies that may degrade the operation of the entire exchange in an unacceptable manner.

However, the most common supervision mechanism in the exchanges work on a *per-call basis*. Thus, as a general principle, the software activities of an exchange manifest during their operation the occurrence of *anomalous situations*. These occurrences are communicated to the maintenance software as syndromes of errors, faults, or failures. The rationale behind call-basis supervision is that of detecting any anomaly during the first call affected by its occurrence. By doing so, at most one call attempt will be disrupted by a fault in the exchange. As long as the occurrence rate of the anomalies stays under reasonable thresholds and if they do not have devastating effects on the exchange operation, this approach is both correct and effective. However, faults that are very critical for the exchange's integrity are handled by continuous supervision.

A form of call-basis supervision is implemented by means of a set of administrable counters in which the ordinary sequences of call and signaling processing sum the frequency of the typical events that they observe. As an example, we have already mentioned the case (see Section 2.1.3) of the physical and frame layers of common signaling channels and ISDN lines. The counters used in these cases are provided with administrable threshold variables, such that, as long as the counters stay within the boundaries stated by their thresholds, nothing occurs. However, when a counter goes beyond its limits, an anomalous syndrome is generated to first-level diagnosis.

The possibility for an exchange device to remain fault-free for a time T, is an exponential function decreasing with T; this means that it is a function like $Y = \exp(-aT)$. This fact means that a frequently used exchange device has a very low probability of being found faulty each time it is used. Conversely, devices that are seldom used may show substantial probability of being faulty when they are needed. Because of this situation, exchange devices that are not continuously used are also checked on a time-basis manner. To do so, time-basis routines are planned and started by specific maintenance activities in the application software.

As a general rule, the exchange maintenance software includes several test procedures, which are very analytical and detailed, used to check the actual exchange behavior from different standpoints. These test routines may be individually started by specific operator commands and can be executed according to planned, repetitive sequences, which are themselves defined and started by operator commands. In any case, the test sequences are executed at low priority, during low traffic situations, and are even designed to cover anomalies that are not detected by continuous and call-basis supervision. These anomalies, by themselves, do not cause devastating effects in the exchange. However, if they accumulate in number, such anomalies may cause complex situations that can make the exchange operation troublesome.

First-level diagnosis, part of the maintenance software, involves identifying and locating the faults in a replicated subsystem, which can be removed from service while a replica takes its place so that service can be restored. First-level diagnosis bases its operation on syndromes received from the supervision. The diagnostics can start verification procedures similar to those included in the time-basis routines, even if necessarily simplified for a quick conclusion. As well as the aspect related to verification, first-level diagnosis includes configuration activities similar to those already seen in Section 2.5.1 for the operation functions. However, unlike the latter case, its triggering mechanism for parameter modifications does not come from operator requests, but rather from the detection of anomalies.

First-level diagnosis is often called fault isolation and service restoration. A faulty device, after it has been put out of service, will need to be repaired and this requires the intervention of human operators. Second-level diagnosis deals with the analysis of a faulty subsystem after it has been taken out of service. The purpose of these procedures, often simply called "diagnostics," is to repair the failed subsystem by identifying which plug-in unit needs to be replaced. Second-level diagnosis consists of a set of *verification procedures* for the analysis of the actual behavior of devices still out of service. These procedures are started by operator commands. Their result is an indication stating whether the analyzed functional block is working correctly. In the negative case, the procedures also give an indication of possible residual faults and their location.

For the implementation of their maintenance function, switching exchanges also can be provided with specific test devices, which are operated in the exchange common control by specialized application activities, capable of requesting to the test devices the implementation of specific functional sequences. The same activities receive results and responses from the devices, and are started either by operator requests or following specific solicitations generated by the above-mentioned verification procedures.

2.5.3 Person-Machine Interactions

The implementation of operation and maintenance activities and semiautomatic calls require considerable intervention of human operators via a multiplicity of operator positions. The management of these positions constitutes a further set of features, which must be provided by the exchange software. In this case, needs and problems may be organized according to the following three points:

A   The software must give the operators a *person-machine dialogue,* which must be, as much as possible, effective and immediate with respect to human needs.
B   The software, in addition to guiding the person in his or her operations, must ensure that these operations are correct and consistent with respect to the current exchange realities.
C   The software must deliver to the relevant operators, in a timely manner, the information specifying each meaningful event in the exchange.

Point *A* and part of point *B* are typically implemented by means of person-machine languages designed to comply with both the needs of computers and the psychological peculiarities of human beings. Correctness and consistency of operator requests are handled by the operation software and, depending on the case, by the activities dealing with semiautomatic traffic. At the CCITT level, a specific person-machine language has been normalized by standard, which is suitable for the operation and maintenance of the exchanges. However, no standard is yet available for semiautomatic traffic. The consistency problems stated in point *B* strongly depend on the way in which the software is structured and cannot be easily normalized or made standard.

Operator positions can be connected to an exchange either locally or through data lines. *On each data line, more than one operator position can be linked to an exchange. This can be done on both dedicated and switched lines.* This means that, according to the case, each operator position interfaces the exchange software, either through a driver used for its terminal or a data line driver. These drivers interface specific application activities dealing with the person-machine dialogue.

When an operator wants to start some activity at a position, he or she sets up interactively with the exchange software, a request for a new *session.* Within this procedure, the operator is typically requested to provide a password for identification purposes. He or she will also provide some parameters to specify the kind of activity he or she wants to perform. If the exchange software finds the data received by the exchange to be consistent, the requested session opens. In any other case, it is refused. Once in session, an operator position interacts with the exchange software according to procedures consistent with the person-machine language accepted by the exchange. The dialogue is structured into *sequences,* each of which, if correctly treated, allows the operator to give the exchange software a well defined request, complete with all the relevant parameters. The exchange software must verify if the request is among those for which the operator is authorized. If not, the exchange software refuses the request; otherwise, it is passed to the relevant software activity, which verifies its consistency and, in the positive case, executes it.

For each operator in session, a queue must be provided, where the messages are placed from exchange software to operator position. Each operator in session must be informed about the conditions of its queue. This permits him or her to request from the software the messages in waiting.

As a further feature, the exchange software must keep a set of *warning and state variables* (typically a few hundred of them) by means of which the operating conditions of the physical and logical devices in the exchange are coded. These variables, forming a *warning and state table,* are automatically updated as part of the normal activities of the *operation software* (see Section 2.5.4). Each operator position in session, by means of a specific operator command, will be capable of specifying to the software exchange the set in the universe of state variables about which he or she wants to be updated. The software exchange consequently operates by showing the current values of the chosen state variables on the operator's screen. All these features must be provided for the operator activities related to the exchange operation and maintenance, and for those dealing with semiautomatic traffic.

Following a period of inactivity that is too long for an operator or after a specific command from his or her position, the software exchange closes the relevant session.

2.5.4 Operation Software

In the previous sections, we have already depicted how signaling and call processing activities also include operation functions. The points expounded on this subject are summarized:

- Each switching activity, during the implementation of its own functions, sums, using specific variables, the occurrences of specific events (either anomalous or not) meaningful for evaluating the actual operating modes of telephone processes. These variables are started, transferred, reset, and made available on the basis of solicitations from application activities dedicated to the operation of the entire exchange that constitute the exchange *operation software.*
- Beside the *ordinary tracking* functions, stated in the previous point, the same switching activities may carry out sequences that generate a more detailed tracking of what they are doing. These sequences require larger amounts of CPU time and are necessary only in special circumstances. Therefore, they are started only when needed for specific physical or logical entities of the exchange. This is done by means of solicitations from the *operation software* to which they give responses and results. (Similar solicitations are used to suspend the functions when no longer relevant.)
- The *tickets* of completed calls to be individually charged are given to the operation software to implement detailed documentation services to the subscribers.

- When individual metering counters are used to charge the subscribers, these counters are updated during call processing and are made available to the operation software upon solicitations generated by it.
- In order to modify configuration parameters that affect only one activity, such an activity implements the updating based on a solicitation received from the operation software, to which it gives responses and replies.
- Modifications of configuration parameters affecting several activities, are *booked* by the operation software at the concerned activities. Each of these activities, when one reaches a condition in which it can accept that parameter change, sends a reply to the operation software. This software, once it has obtained all the replies, modifies the parameters and then communicates to the same activities the resulting update.
- For implementing traffic measurements, the switching activities use counters, as discussed in the first point of this list.
- The exchange software includes a warning and state table in which the operating conditions of both the devices that make up the exchange and its logical resource are coded. This table is updated by the operation software based on the solicitations received from the maintenance software and by modification procedures requested by operator positions and remote processors.

The operation software (see Figure 2.44) interacts closely with the *operation capabilities* included as part of the switching software and the drivers of the common control peripherals. The operation software carries out its functions based on solicitations received by local and remote operator positions, remote processors emulating the activities of operators, and the maintenance software (Figure 2.45). For each solicitation, the operation software carries out a sequence of *transactions*. Examples of transactions are parameter modifications, measurement executions, counter transfers, *et cetera*. Some transactions are short (e.g., some parameter modifications), others may last longer (e.g., measurements). The operation software may concurrently accept and start several transactions, provided that they are compatible with one another. Each kind of transaction finds its counterpart in a functional segment of the operation software. This means that (see Figure 2.46) the operation software includes a library of functional segments implementing the different transactions. At any given moment, the operation software will start and execute up to a maximum number of different transactions. The executed transactions are based on the received requests and after having checked that each new transaction is compatible with those already in progress. Each transaction may finish naturally, by itself, or abruptly because of specific interruption requests from operators or remote processors provided with interruption capabilities. At any given moment, an operation software functional segment can also be activated on more transactions. For this to happen, the operation software receives more solicitations to carry out the same function on different exchange entities.

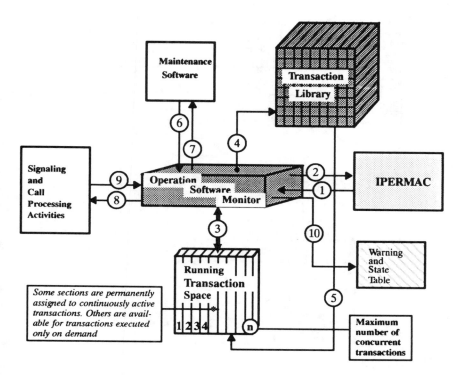

① Requests from the operator positions to start or stop transactions

② Responses to the operator positions

③ Inputs or Outputs to or from running transactions

④ Loading requests from transaction library to transaction space (with all the relevant parameters to be added to each transaction to be transferred).

⑤ Transaction code, already configured, being transferred to its assigned transaction space section.

⑥ Requests from the maintenance software to start/stop transactions (Typically: change in the operating conditions of exchange devices).

⑦ Responses to the maintenance software

⑧ Directives to the signaling & call processing software

⑨ Responses from the signaling & call processing software

⑩ Updating of state variables describing the current conditions of the exchange modules for operators and remote processors.

Figure 2.44 Operation software structure.

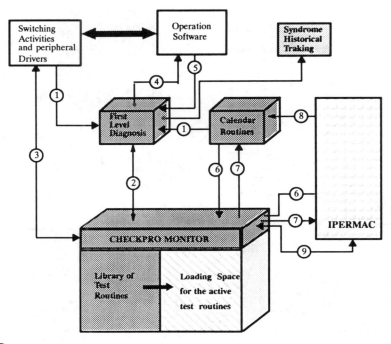

① Anomalous syndromes

② Requests to run test routines and related responses

③ Back and forward messages exchanged by the test routines in progress

④ Requests to update state variables of devices and resources, at values such as: "in service, out of service, in line, under, test etc."

⑤ Responses related to the requests of point ④

⑥ Request to start or stop test routines

⑦ Responses from test routines in progress

⑧ Updating of the parameters describing the schedules for the test routines to be executed on a calendar basis

⑨ Requests to test and put in service repaired or new units (with related responses)

Figure 2.45 Maintenance software structure.

In addition to problems related to the compatibility among different simultaneous transactions, the amount of work that the operation software can run in the exchange has an upper bound, which depends on the processing capabilities of the exchange common control. Similarly to the case of any application activity structured around parallel entities to run *on demand,* the operation software also includes a *functional segment,* which organizes all the work, like an orchestra conductor, and monitors the succession of requests originated by operator positions or remote processors.

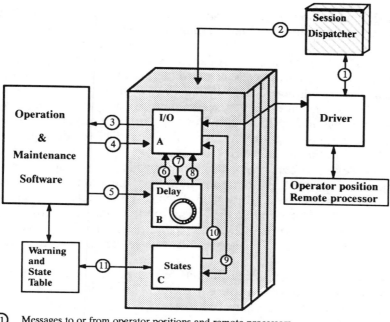

(1) Messages to or from operator positions and remote processors
(2) Assignment or Release of sessions to or from operator postions
(3) Requests to the operation and maintenance software
(4) Responses in session from the operation and maintenance software
(5) Delayed responses from the operation and maintenance software
(6) Indication of some or no message in waiting
(7) Requests related to the queue of messages in waiting
(8) Sorted messages in waiting
(9) Requests for changes of the state variables to be selected
(10) Updatings on the selected state variables
(11) Polling the selected exchange state variables

Figure 2.46 Structure of the software for person-machine interactions.

2.5.5 *Maintenance Software*

Similarly to the case of the operation software, the switching application activities also implement substantial maintenance features. In fact,

- Signaling pre- and postprocessing activities and peripheral drivers also deal with the alarm signals generated by their peripheral hardware and their exchange terminations, and send *bad working syndromes* to the maintenance application activities of the *maintenance software*.

- As a general rule, every switching activity constantly checks for any illegal event and for any inconsistent situation. Whenever any of these happens, in addition to clearing the calls involved, the same activity generates a well documented syndrome to the maintenance software.
- The switching activities, while trackings and summing their *counter variables,* check for any out of threshold and for any downtime or "time-out" to occur. These conditions are automatically communicated as well documented alarm syndromes to the maintenance software.
- Signaling postprocessing activities accept maintenance solicitations for terminations and peripheral units, and translate them into testing commands to these devices. The results are then detected by the relevant preprocessing activities, which filter and forward them to the maintenance software.
- Call processing activities include, as well as the sequences necessary for the useful traffic, *test call sequences,* which, when started, give their results to the maintenance software.
- Test calls also use measurement and testing devices connected to special terminations. These devices reply to their drivers in the application software. Each of these drivers is equivalent to a special signal pre- or postprocessing activity, which interacts with the test calls.

The maintenance software tends to be structured into the following three main blocks (see Figure 2.45):

- A functional segment that implements the first-level diagnosis (FLD);
- An application activity that includes the testing procedures (CHECKPRO);
- An application activity for the execution of tests on a calendar basis (CALENDAR).

The segment FLD is the entity where any anomalous syndrome generated by the remaining software is sent. FLD filters and processes them, one after the other, and takes the proper decision in any circumstance.

The possible results of FLD can be grouped into the following options:

- Requests to CHECKPRO to execute one or more testing procedures, to obtain more detailed information for FLD about the ongoing faulty condition;
- Requests to the operation software to start one or more transactions to put faulty elements out of service and to switch over their traffic to available standby units;
- Requests to the operation software to update the warning and state variables describing the actual conditions of the physical and logical resources available in the exchange;
- Based on the results obtained from CHECKPRO and FLD, storage of messages into historical files or on operator queues to describe events and circumstances related to the detected faults.

The application activity CHECKPRO is a library of functional segments, basically one segment for each *test procedure*. As in the case of the operation software transactions, more CHECKPRO procedures may be executed concurrently. Each procedure consists of an algorithmic sequence of requests to execute *specific tests*. The start of the procedures and the restoration of their results are managed by a monitor that is another functional segment of CHECKPRO. This monitor segment operates on the basis of:

- Requests of further tests coming from FLD;
- Requests for second-level diagnosis originated at operator positions;
- Requests from CALENDAR.

The result of each test is given by the CHECKPRO monitor to the entity that generated the requests. As in the case of the operation software transactions, the CHECKPRO monitor refuses to start concurrent testing procedures that are incompatible with each other and limits the maximum number of these sequences, according to the processing capability of common control.

CALENDAR carries out the following functions:

- Following specific indications from operator positions and remote processors emulating them, it puts in or takes out of a calendar schedule each test that must be done on a regular basis. This is done together with the timing according to which it needs to be executed;
- It requests CHECKPRO to execute procedures already scheduled when their time arrives;
- It transfers the results of the procedures on a storage support, such as printers and files, which are accessible to maintenance operators.

Some of the testing procedures included in CHECKPRO are used to operate test devices connected to the exchange (see Section 1.1.7). These testing procedures may be activated by either operator commands or called by sequences in CALENDAR. At least in principle, the results are logged onto files that operators can access and search via commands that trigger specific transactions in the operation software. Also, in this case, if situations arise that require an operator to be alerted, a warning condition occurs in the warning and state table of the exchange.

2.5.6 *Person-Machine Interaction Software*

To implement the interactions between persons and machines, the application system must include a specific activity (IPERMAC). IPERMAC interfaces, on one side, the operation and maintenance software and the semiautomatic call processing; on the other side, IPERMAC interacts with the operator position drivers and the pre- or postprocessing activities of the data lines (this for remote processors or remote operator positions). Typically, in switching exchanges dealing with semiautomatic traffic, there are two IPERMACs: one for the operation and maintenance positions, the other for semiautomatic call processing. These two IPER-MACs tend to be different because of the dissimilar requirements associated with operation and maintenance and the treatment of semiautomatic calls. However, because of their similarities, the two kinds of IPERMAC are nonetheless considered together.

As a first element, IPERMAC includes (see Figure 2.46) a functional segment for *session assignment*. This segment, based on the initial messages coming from the positions, does the following:

- It requests possible passwords from the operators;
- It refuses session requests which are incompatible with common control capabilities;
- It assigns, whenever possible, a session to any operator position at which a request has been originated.

For IPERMAC, the *sessions* are seen as logical resources. In the switching systems, two possible alternatives can be found:

- A specific proprietary session is associated with each operator position, which may be either dormant or active;
- The sessions are a global resource and are distributed to the operators when required. A session associated with an operator position is removed from it and made available for other assignments, following either an operator command or an extended period of inactivity.

In addition to the session assignment segment, IPERMAC includes three other segments, which it uses to implement the sessions. These three elements are instanced with the sessions of IPERMAC. They are (see Figure 2.46):

A   *Person-machine dialogue management*;
B   *Delayed message delivery from software to operators*;
C   *Updating messages to the operators*.

Segment A includes acceptance, analysis, and verification of operator requests, and transforms any accepted request into solicitations to the remaining software. These solicitations, according to the case, go to:

- The operation software;
- The maintenance software;
- Segment B;
- Segment C.

The procedures developed between operators and the remaining software via segment A are typically interactive. Segment A implements its side of the communication protocol; it verifies that everything is correct and closes the relevant session if anything does not occur properly. The result of a person-machine interaction may be the initialization of functional segments that do not give the operator immediate answers. In this case, the operator receives only an acknowledgement stating that the relevant software segment has started.

The results of the actions of the software, when they are not given to the operator positions during communication between an operator and the exchange, are placed in a waiting queue, which is handled by segment B. Each time an operator's queue is not empty, segment B relates this fact to the relevant position via segment A. Then, following directives received by the operator via segment A, segment B sends the relevant operator positions the messages in waiting. According to the operator indications, these messages can either be cancelled from the queue or kept there.

IPERMAC operates on the warning and state table, which analytically states the operating conditions of the exchange, as determined by first-level diagnosis and the operation software. On these variables, an administrable standard subset S_V is defined. Typically, each operator position is continuously updated about the current value of S_V. As an alternative, each operator in session may require segment C to give to the operator position, instead of S_V, another personalized subset of the state variables. On this subject, the operation position is provided with commands for choosing the personalized subset. These commands, received by segment A, are sent to segment C, which accordingly updates the operator positions by scanning the current values of the warning and state table.

2.5.7   Measuring Procedures

Measures of the activities of an exchange may be implemented according to two general methods: (1) based on the use of (enriched) call tickets and (2) spreading *counter variables* and *state variables* throughout the exchange software. The tickets (call records) of both successful calls and unsuccessful call attempts may include any information pertaining to the circumstances under which each call or call attempt occurred. By doing so, and through postprocessing the tickets, to measure most of the exchange activities is possible. Call tickets are very efficient for measuring items related to the traffic flowing through the exchange. They can be used to measure both traffic load (in erlangs) and call attempts of any kind. These items can be evaluated globally and in terms of traffic streams, incoming from various origins and outgoing to allowed destinations. Tickets may also be used in a more analytical way to track special calls (for example, malicious call attempts, calls on fuzzy lines, *et cetera*) for which larger and fully detailed tickets may be generated. When there would be too many tickets to handle, similarly to studies in demography, the exchange could be programmed to send only suitable samples of tickets, instead of all of them.

The main disadvantage of using call tickets is that they allow for only *postmortem* measures, and do not fill the need to have on-line indications about the traffic being carried. In this respect, counter and state variables are a better choice. State variables are associated with objects such as lines, bundles, directions, destinations, or processors, to state at any moment their actual operating conditions. These variables are meant to be shown on request at the operator positions and to be regularly sampled by the exchange software, which in this way may generate statistics about activity. These statistics permit obtaining traffic and load measures related to lines, bundles, *et cetera*, in a more straightforward manner than by postprocessing a population of call tickets. State variables are obviously set and updated by the exchange software during its normal activities to reflect at any moment the actual situation. Counter variables are associated with events that may happen on specific physical and logical resources of the exchange. For example, a counter variable can be associated with any given line to count the number of seizures occurring on it, another to count the calls that have obtained an answer from the called party, and so on. Also, these variables are continuously updated by the exchange software as part of its current on-line activities. State variables and counter variables are used in the exchange software to be read and, if necessary, reset by special application activities that implement measurement processing.

On this subject, a general reference model has been defined at the international level, and is briefly described here. Counters and state variables are seen as *entities*. An entity is either a type of state variable or a type of counter with a well defined meaning. Each entity, by definition, is associated to a set of *objects* to which it applies. (As an example, the objects of the entity that is *number of occurred seizures in the incoming lines* are the incoming lines of the exchange). Each entity has its objects. Different entities may have objects in common (see Figure 2.47). The logical sum of the objects of all the entities form the *universe of objects* on which measures can be taken. Entities and objects are the first two ingredients of measurement processes based on counters and state variables. The third ingredient is the *timing,* which defines when a measurement must be performed. Examples of timing follow:

- The day 1, 4, 15, and 30 of this month; each day from 4:35 to 7:55 *p.m.* and from 1:00 to 4:00 *a.m.;* during each active period, one measure to cover each and every period of five minutes.
- Every day of the month; from 3:00 *a.m.* to 10:00 *p.m.*; with one measure taken over each 10 consecutive minutes.
- The day 5 of this month; from 00:00 to midnight; with one measure each consecutive 15 minutes.

As should be apparent from the above examples, a timing has three components (see Figure 2.48):

1. The days of the month in which the measure will be done;
2. The hours in each active day during which the measure will occur;
3. The time interval in which the active hours will be divided.

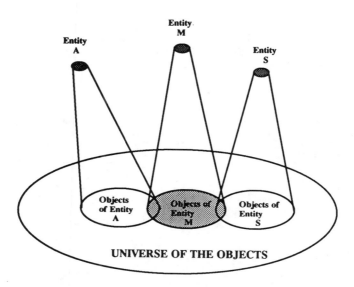

Figure 2.47 Entities and objects.

Hence, a measure will be taken for each time interval in the active hours of the active days.

A *measure type* is defined as the following triplet (see Figure 2.48):

1. A set S_E of entities applying to the *same* objects;
2. A subset $L(S_E)$ of the objects to which all the S_E apply;
3. A timing $T(S_E)$.

The meaning of the measure type is to take the measurements on S_E, applied only to the $L(S_E)$, in the time span set by $T(S_E)$. The result of the measurement process on each consecutive time interval (of 5, 10, 15, . . . minutes) for which an entity is a counter is the increment of that counter over the time interval. The result of the measure of a state variable is the relative frequency for each configuration allowed for the state as sampled during the time interval (e.g., 90% busy, 10% idle).

A *measure* (see Figure 2.49) is defined as any *set of measure types* in which entities and objects may be defined independently of each other, *while the time is chosen identically for all types.* Theoretically, types of measures and measures could be defined independently of the meaning of the chosen entities. Practically, the opposite happens, as a measure is intended to be a reference model to assemble useful and related information into well defined sets of results obtained by the exchange application software. This can be done only through a careful choice of entities and objects assembled in the same measure.

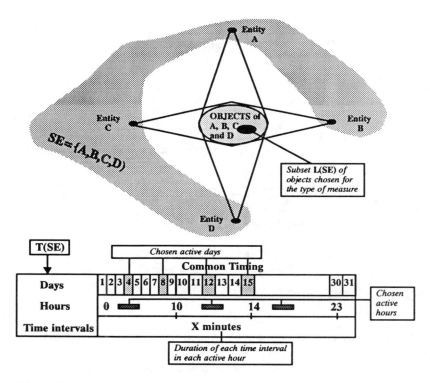

Figure 2.48 Type of measure layout.

CONCLUSIONS

Functions and features analyzed in this chapter are a superset of those in each specific exchange. In comprehensive switching systems, they are implemented, together with the operating system, as a *library* of application programs provided with the relevant data structures. For each type of exchange, a suitable *generic* package is produced from the library by integrating the list of functions applying to that case. Each generic has a typical size ranging from several hundred thousand to over one or two million lines of code.

The universe of the application systems activities can be partitioned into four parts:

1. Signaling pre- or postprocessing;
2. Call processing;
3. Operation software;
4. Maintenance software.

In terms of lines of code, parts (1) and (2) cover less than 25% of the total of a generic; while parts (3) and (4) are over 55%. (The remaining 20% or so is

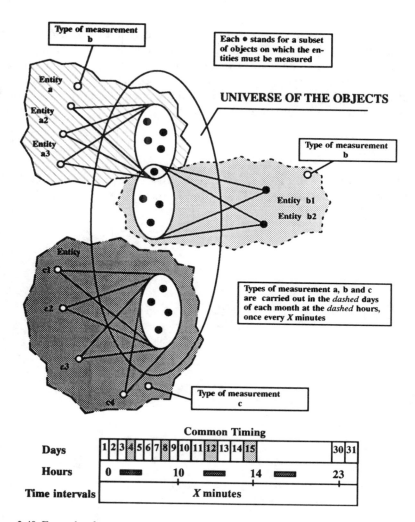

Type of measurement b

Each ● stands for a subset of objects on which the entities must be measured

UNIVERSE OF THE OBJECTS

Entity a

Entity a2

Entity a3

Type of measurement b

Entity b1

Entity b2

Entity c1

c2

c3

c4

Types of measurement a, b and c are carried out in the *dashed* days of each month at the *dashed* hours, once every X minutes

Type of measurement c

Common Timing

Days	1 2 3 4 5 6 7 8 9 10 11 12 13 14 15 ... 30 31	
Hours	0 ▬▬ 10 ▬▬ 14 ▬▬ 23	
Time intervals	X minutes	

Figure 2.49 Example of measure.

due to the operating systems). In terms of CPU time, parts (1) and (2) consume the larger share, especially during peak traffic conditions. However, parts (3) and (4) are mostly operated at a lower priority and may delay functions that can wait some time before being done.

For very large exchanges, the processing load can approach peaks of 1,000,000 call attempts per hour, with peak values of about 30,000 calls in progress at the same time. These values set the needs of the processing capabilities and degrees of parallelism needed in the software of switching exchanges.

The operation and maintenance software, even if it does not produce any useful (i.e., *paid by the subscribers!*) traffic, *per se,* is of great value and importance for the operating companies because it makes possible, and easier, the services offered to the subscribers. Because of this fact, the operation and maintenance software naturally tends to grow in digital switching systems as their use spreads in public telecommunication networks.

Bibliographic Note

In this chapter, we make systematic reference to CCITT Recommendations. The interested reader may refer to the CCITT "Blue Books" (ITU, Geneva, 1989).

Chapter 3
Alternative Architectures of Digital Switching Systems

INTRODUCTION

Note: An analysis, even superficial, of the realities of digital exchanges, shows how the most discriminating factors in their architectures are strongly related to the common control functions: to the ways in which they are organized for the multiplicity of processors in the common control, to the logical and physical architectures adopted for it. However, the switching matrix, although it can be implemented in different ways, shows architectural effects that are becoming rather marginal. This is because of the success of VLSI technologies that are making the design of switching matrices more and more simple. Given this situation, in the sections of this chapter analyze the choices that emerge from the realities of the digital exchanges now available on the market.

The starting point for our considerations is the first generation of electronic exchanges (see Section 3.1). They lead to discussion of the architecture of today's switching exchanges with a monolithic common control. To overcome the potential limits of this architecture, several improvements have been tried on actual systems, along different paths. The main direction in this respect (see Sections 3.2 and 3.3) is the introduction of peripheral processors around the common control and the definition of support computers on which to transfer most of the operation and maintenance functions.

As further steps, we analyze the problems that emphasize the fragmentation of a common control into a network of processors (see Section 3.4). At this point (see Section 3.5), the case of Alcatel's System 12 is shown, which is currently the most dramatic example of an exploded architecture for common controls.

In the later sections, we study alternatives, which naturally lead to compromise approaches between monolithic and exploded common controls, and have been successful in practical communication networks.

In the concluding section, we provide a synthesis for the architectures used in digital switching exchanges.

3.1 Architectures Using a Monolithic and Centralized Common Control

The quasibipolar model of switching exchanges, discussed in Chapter 1, suggests by itself a radical systemic approach, according to which:

- Circuit switching is done by the matrix;
- Exchange terminations (with the exception of those for data and common signaling channels) are plain electrical interfaces that separate useful signals from signaling;
- The control functions are totally centralized in a monolithic control;
- Multifrequency registers are frequency transducers with a minimum of processing capability;
- Exchange terminations for data links, common signaling channels, and data multiregisters, as well as auxiliary units, are considered as external to the exchange.

This approach, even with some correctives, had been the one followed in the switching systems of the 1960s and 1970s, when microprocessors were still to be invented.

In those years, in the absence of microprocessors, the only practical approach in the development of switching systems consisted of designing *stored program control* (SPC) exchanges (see Figure 3.1) with a monolithic common control, where a single piece of software implemented everything. The logical functions that were absolutely necessary in the exchange terminations, to avoid intrinsically inefficient uses of the centralized control, were implemented by means of logical circuits, either combinatorial or sequential. This, because of the difficulties implicit in the implementation of complex functions using only circuitry, brought about greatly reduced distributed intelligence in the exchange terminations and markers. Examples of such minimal logic levels are:

- Treatment of associated signaling for PCM lines in channel 16;
- Internal resource management and supervision of the switching matrix by means of a marker;
- Line signal processing on subscriber loops.

The idea of implementing every supervisory and control capability of the exchange by means of a software package executed by one CPU, still provides substantial flexibility. In fact—although writing, modifying, and testing such soft-

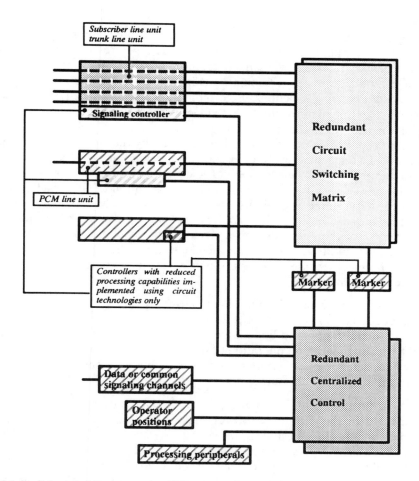

Figure 3.1 Basic layout of first-generation SPC switching exchanges.

ware is not an inexpensive and simple activity—software, once developed and tested, can be replicated at relatively marginal costs. This fact makes intrinsically competitive approaches that reduce a switching exchange to a few simple and repetitive circuit elements.

By doing so, the recurring costs of an exchange, typically associated to the production, testing, and maintenance of its hardware, may be dramatically reduced. In this way, the overall costs of an exchange produced and sold in numbers large enough to amortize software and hardware design costs, may be brought down.

As already mentioned, the technology available in the 1960s and 1970s made convenient the development of an SPC system in which only the common control was a stored program device (i.e., a redundant CPU with its software). The remaining part of the exchange was implemented by using combinatorial sequential logical circuits. This brought about exchange architectures with very simple terminations and registers. At the same time, the needs for features requiring complex exchange terminations and common signaling channels were not mature; neither were packet-switching functions. Like today, auxiliary devices (few and simple) were seen as entities external to the exchange. In those years, the terms of reference were the needs of telecommunication networks related to services, dimensions, and traffic capabilities that could be implemented by means of electromechanical exchanges. In this situation, SPC exchanges appeared as significantly innovative and competitive for the development of telecommunication networks.

The most difficult problems in the implementation of SPC exchanges were related to the required availability of centralized controls. Obviously, a control which fails or must be stopped to modify or reload its software means interrupting the operation of the entire exchange. Unlike what happens, for instance, in the case of electric utilities, a telephone exchange cannot be stopped, even for short periods of time, because this means putting all its subscribers out of service. Because of this situation:

- A centralized control is to be characterized by outage probabilities that must appear statistically marginal over periods of decades;
- Any modification and updating of software, including alterations related to the maintenance and operation of the entire exchange must be done on-line, while the exchange continues to operate, without any substantial interruption in the services to the subscribers.

As a further point, the multiplicity and complexity of the functions to be implemented by the exchange software make detecting and removing every programming error a difficult and laborious task. At the same time, any problem not already fixed in the software may cause, at the most, situations that the common control is able to overcome, practically without interruption of service. This means that the software of a centralized control must be highly reliable against residual faults it may still contain (and that it typically also contains after having operated several months on different exchanges!).

The reliability and availability prerequisites fundamentally require that the centralized control be implemented in a redundant way. This is because the reliability of the components and manufacturing processes (such as mounting, soldering) makes nonredundant processors insufficiently reliable.

Studies, experiments, and implementation since the late 1960s came to conclude that the most efficient redundant structure for centralized controls consists of duplicate processors, which, from the application standpoint, appear virtually as if they were a single device.

Basically, the sections of such a control, when both are active, either do the same things in parallel or operate according to methodologies whereby one carries out the useful application processes, while the other operates in standby, ready to switch-over following any failure in the first unit. The redundant architectures adopted for centralized controls of telephone exchanges are discussed in Chapter 4. Hereafter, we emphasize the following three points:

- The definition of these architectures is one of the critical points in the description of SPC exchanges;
- The chosen solutions, defined for the specific needs of the exchanges, offer an original contribution of telecommunication techniques to the science of information processing, but find limited usage in other fields;
- The same redundant architectures also have been employed in the successive exchange architectures defined since the late 1970s and still in use today.

Obviously, the processing load of a centralized common control depends on the size of the exchange. More precisely, the needed processing capacity depends, above all, on the number of call attempts per unit of time with which the machine must deal during peak traffic hours. The capacity also depends on the processing load associated with the implementation of operation and maintenance. With the growing size of the exchange, the practice of switching systems has shown how the processing capabilities of centralized controls quickly reach a bottleneck. This means that if one wants to maintain the radical choice of keeping most of the exchange logic in a single centralized control, it quickly becomes too fast and too costly. As a consequence, structures with a centralized control but without intelligent periphery, while interesting for switches of 100–1000 lines, lose their practical value for larger exchanges. This requires the adoption of different architectures defined since the late 1970s and made possible by the dramatic evolution of microprocessors, the reduction in the cost of memory, and the rapid growth of their capacities.

3.2 Regionalized Common Controls

The coming of microprocessors has made possible, since the 1970s, the implementation of exchange terminations, markers, and other peripheral units characterized by substantial levels or processing capabilities by means of software techniques. This implementation has made possible the decentralization (or *regionalization*) of some of the activities of the exchange software, thereby reducing the processing load of central control. A *regional processor* hence is one that handles the full complement of functions for a subset of physical entities (e.g., registers).

Note: From this point onward, the term *centralized control* will no longer be used to emphasize that not all the processing capability in the exchange is in the central part of the common control. Instead, the term *central control* is used.

Conversely, the term *common control* continues to be used for the totality of processing functions in an exchange.

Referring to Chapter 2, the software activities that may be conveniently regionalized from the central control are synthesized in the following paragraphs A to G (see Figure 3.2).

A. Analog Line Units (Subscriber Loops and Junctions)

On each line unit, the preprocessing and postprocessing activities of the telephone signaling can be regionalized. In this way, the central control sees each line as transceiving only complete telephone signals, each consisting of well defined messages instead of binary variations of signaling leads. On the interface between control and unit, in addition to the telephone signals, directives and responses of operation and maintenance are sent in both directions.

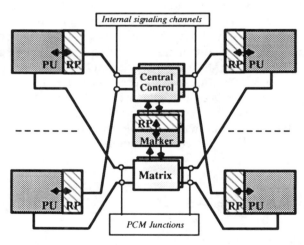

PU : Peripheral unit including a set of

* Subscriber (analog) lines
* Analog trunk lines
* ISDN basic rate interfaces
* ISDN primary rate interfaces
* PCM lines
* Data multifrequency register
* Auxiliary devices

RP : Regional processor dedicated to the traffic and O&M processing of its PU

Figure 3.2 General layout of an exchange with regionalized central control.

B. PCM Terminations

By using microprocessors, PCM terminations may implement by way of software the processing related to the insertion or extraction of the association signaling, the test routines on both line and unit, and the preprocessing and postprocessing activities of telephone signaling. Also, in this case, the interface between unit and central control is based on message-oriented protocols.

C. Common Signaling Channels, Data Links, Data Registers

Physical and frame layers may be easily regionalized. This allows for the central control to interface its data links and common signaling channels by interfaces that are uniform for the entire exchange.

D. ISDN Line Units

Activities related to the frame, physical, and network layers of the D channels can be regionalized. An ISDN line unit may interwork with the central control by transceiving:

- data packets to or from the packet-switching activities,
- telephone messages from or to INLINE or OUTLINE,
- directives or responses from or to the operation and maintenance activities.

E. Markers

Each marker may be provided with the processing capabilities necessary for the central control to see matrix and marker as a virtual matrix from which it only asks for connections and disconnections between any pair of terminations. From the same virtual matrix, the central control obtains positive and negative acknowledgements. The virtual matrix contains supervision procedures and first-level diagnosis for the verification of the proper behavior of matrix and marker. In this way, the maintenance and operation activities related to circuit switching, which are still in the central control, are limited to repair procedures and person-machine interactions. Repair procedures are still in the central control because they require the use of mass memory devices, which are not convenient to attach to the matrix. Person-machine interactions are still in the central control because they are implemented by means of centralized software activities, which apply to any exchange device.

F. Multifrequency Registers

The processors in the frequency multiregisters may be virtualized so that the central control sends to each register and receives from it messages requesting or stating the occurrence of frequency combinations. These processors will also perform tests on the registers, started by requests from central control and providing it with their responses.

G. Operator Positions

An intelligent interface with central control transforms an operator position into an intelligent terminal, which may include most of the functions related to person-machine interactions. Specifically, an intelligent operator terminal may have the following features:

- Analysis of each person-machine interaction, including the identification of the action requested by the operator. This action is sent to the central control for further verification;
- Reception from central control of anomalous syndromes and other kinds of messages for the operator. These messages can be formatted and put in waiting, available for the operator to read. Interactive procedures can also be made available, through which an operator knows when messages are ready to be read, sort them on the screen or printer, and delete them from the list of waiting messages.

An intelligent operator position is seen by the common control as an external processor with which it interacts according to message protocols that may be set at the system level. Common control architectures in which peripheral activities are regionalized are rather common today for electronic switching systems. An example in this respect is System AXE from L.M. Ericsson.

The potential advantage of structures with monolithic central control is that one (redundant) processor handles all the logical and physical resources of the exchange, and can optimize, on a moment by moment basis, the use of the exchange's resources. Examples of resources which can be handled efficiently by a central control structure are trunk groups, common signaling channels, and multifrequency registers.

The potential disadvantage of centralized control architectures remains the fact that, with the growth of traffic, number of subscriber lines, number of trunks, operation and maintenance features, and additional services to the subscriber, the processing capabilities of the central control remain a limited resource continuously in danger of becoming saturated. This typically implies the use of special processors, which are difficult to derive from the components normally available for data processing applications. Nonetheless, only by exclusively using these components can one enjoy the cost-performance advantages due to the tremendous volume of production associated with data processing applications.

Another potential disadvantage in architectures using central controls is that, by doing so, the central control becomes a substantially monolithic block that must be tailored to the maximum capacity of the exchange. This means that the cost structure in exchanges with a central control is a win or lose situation, with substantially zero-sum costs. This may affect the competitiveness of these exchanges in the lower ranges of their field of application.

Note: In any case, advantages and disadvantages, as quoted so far, are potential but not absolute. This is because, when expressed in terms of money for a specific situation or system, an advantage or disadvantage can become marginal against other factors that are not related to architectural features, but to other elements, industrial or commercial in nature.

In an architecture using a regionalized central control, such a control must be duplicated. This is not necessarily so for processors used in trunk line units, multiregisters, common signaling channels, data links, and other auxiliary units. In these cases, in fact, a fault in any of these elements can be superseded by the central control by making use of the redundancies available in the exchange equipment. Thus, when a register, PCM line, trunk line, or common signaling channel fails, the exchange has other units available, equivalent to those in trouble, on which the traffic can be rerouted. For the case of subscriber lines, the redundancies of the relevant processors may be either necessary or not, depending on the number of lines connected to the same unit, which therefore are out of service when the microprocessor fails.

3.3 Service Computers

The operation and maintenance functions are structured into two application levels:

- In the software dealing with signaling and call processing;
- In specific *operation and maintenance activities,* including the interactions with operators and remote processors that emulate the behavior of operators.

As has already been noted in Chapter 2, both the application activities that can be regionalized and those related to call processing:

- Include tracking variables and counters, the content of which is sent to the operation and maintenance activities;
- Accept, by the operation and maintenance activities, requests to start sequences for tracking specific aspects or segments of the traffic handled by the exchange. Also, the results of these trackings are then transferred to the operation and maintenance activities;

- Execute diagnostic sequences, substantially similar to those related to the operation of real traffic, but complemented by the verification of the operating conditions of devices of the exchange. These sequences are started by and give their results to the operation and maintenance activities.

Except for the implementation of first-level diagnosis, in every respect, the operation and maintenance activities operate as an intelligent interface between the operators and the signaling and call processing software. In one direction, they synthesize the relevant results for the human operators; in the other direction, they actuate directives and requests originated by the same operators or remote processors emulating their behavior. Therefore, most of the operation and maintenance activities must be structured according to real-time multiprocessing environments consistent with the speed requirements of human beings (i.e., with answer times of seconds!). These needs are less pressing then those of call and signal processing for which required response times range from a few tens to a few hundreds of milliseconds. Moreover, the number of concurrent signaling and call processing actions that must be active may grow to several thousands as this number is related to the calls simultaneously in progress. However, with the same traffic, the number of concurrent transactions related to operation and maintenance are on the order of 10 to 30. Conversely, the operation and maintenance activities use mass memories and need operating system features typical of electronic data processing (EDP) administrative environments. These needs generally are not found in call and signaling processing.

As opposed to the case of signal and call processing, with the exception of first-level diagnosis, operation and maintenance activities are substantially similar to real-time multiprogramming applications typical in business and administrative informatics. In this domain, because of its tremendous market dimensions, processing systems are available which are very competitive in terms of both performance and cost. This makes significant the idea of moving the operation and maintenance activities, including those related to the operator positions, but not first-level diagnosis, from the central control to a specialized *service computer* (SC) (see Figure 3.3). At the interface between service computer and central control, an information flow occurs, which includes directives and responses in both directions. Mass memories, processing peripherals, operator positions, and data lines to other processing centers are connected to the service computer. The considerations about regionalization also apply to this case. Therefore, intelligent interfaces can be connected to a service computer, similarly to those assumed for the case of central control. A service computer may be located either in the same place as the exchange, or at a remote site. The service computer also may be connected to more than one exchange to be able to do the following:

- Share its costs among the maximum possible number of subscribers;

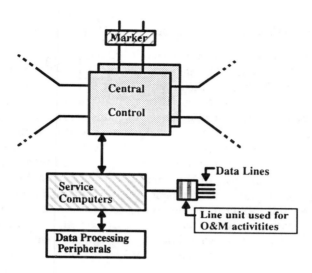

Figure 3.3 General layout for the use of service computers.

- Implement operation and maintenance functions not only for an exchange, but also in a more integrated way, for all exchanges in a given territorial unit of a public telecommunication network.

The functions that can be typically put on a service computer can also be interrupted for periods of up to a few hours without affecting the actual traffic flow. In fact, service computer outages result in only momentary interruptions of the operation and maintenance activities and the possible repair of faults already isolated by first-level diagnosis. *Note*: This must remain within the central control. With the use of a service computer, central control should also maintain the capability to keep the data necessary for charging the subscribers, even when the service computer is momentarily out of service.

This is meaningful because it permits the implementation of the functions in the service computer without using tightly redundant structures, but rather only using products available for business and administrative informatics. These systems, among other things are also provided with software packages that permit easier modification in the operation and maintenance activities, especially those related to a more efficient use of database management systems.

The concept of the service computer as an architectural choice in the implementation of telephone exchanges has been verified with the success of several variants in many switching systems. An example is the support computer adopted in the UT Line from ITALTEL.

3.4 Segmentation of the Central Control

The considerations of the previous two sections, demonstrate a general approach, according to which the criticality of the processing capabilities in telephone exchanges is solved by thinking of the common control as a network of processors that share the functions to be executed or the devices to be controlled. This is done with the service computer that carries out most of the operation and maintenance activities and a number of regional processors, each dedicated to specific functions executed for a set of peripheral devices. At this point, the question arises as to how far such a fragmentation process of the common control can be pursued. The key factor in the analysis of this question is related to the communication mechanisms among programs. Programs in the same processor communicate with each other by reading and writing commonly accessible variables and by sending, via the memory, data messages to each other. Conversely, programs in different processors communicate with each other only through messages sent via the input-output systems of the processors. This second mechanism consumes much more CPU time than the first one.

Let (see Figure 3.4):

- C_i be the average CPU time that a program i spends communicating with the other programs while performing its activities associated with an amount of traffic assumed as the *reference amount of traffic* during a given time T_R;
- P_i be the *overall* CPU time consumed by the same program i under the same circumstances;
- C_i and P_i be evaluated with *all* the programs i placed in the *same* central control.

By summing the P_i of all the programs in the central control, a $P = \Sigma_i P_i$ is obtained, which can be, at most, equal to the time T_R in which the reference amount of traffic occurs (i.e., $P < T_r$). The traffic handled by the exchange in the amount of time T_R can grow to the point where P approaches T_R. With the central control exploded into n processors, each P_i grows to a new value P_i^* because C_i, (which is part of P_i as a general rule) grows substantially. Consequently, P also increases up to P^* (i.e., $P < P^*$ with the same amount of traffic occurring during T_R). In this case, however, the amount of traffic flowing through the exchange during the time T_R can grow to the point where P^* approaches $n \times T_R$. This is because the processing load can be shared among n processors. At this point, a critical need is for P^*/P to remain substantially lower than n. Otherwise, the cost paid by exploding the central control into n processors is verified by the processing overload due to the growth of the C_i. Because of the close interactions among the software elements of a common control, for the inequality $P^*/P \ll n$ to hold is by no means given.

The inequality does hold when moving part of the processing load to the regional processors and with the introduction of the service computer. For it also

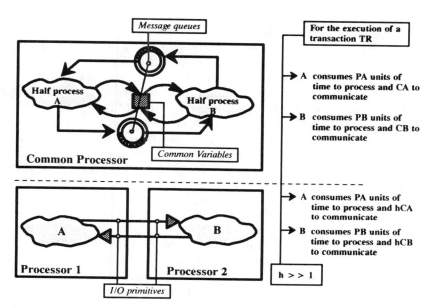

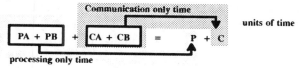

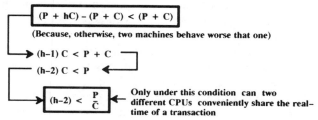

Figure 3.4 Comparison between the processing capabilities of one and two CPUs for the real-time execution of a transaction.

to hold following further subdivisions of the central control into several processors, one needs to assign very carefully to each processor its share of application activities. Such an assignment must be done to provide, at least for most of the programs i, that $(C_i^* - C_i)/P_i$ remains substantially lower than 1 (i.e., that the processing overload, due to the heavier communication processes in the programs, remains marginal as compared with the original processing loads of these programs).

Generally, the problem of successfully dividing a central control into n processors is equivalent to partitioning the exchange software into n sets, which, in order to communicate with each other, spend still marginal amounts of time compared with their overall processing time. Hence, when each set of programs is placed in a different processor, the CPU time spent by the programs to communicate with each other, although higher, is still reasonable and does not jeopardize the increase in processing capabilities that can be obtained by splitting a central control into n processors.

As a further objective, partitioning must also provide that the overall time spent waiting for the communication processes among programs to occur does not make the common control lag behind the calls with which it is dealing. This point is not a trivial one because the CPU time can remain reasonably limited, while the overall communication time (comprising CPU time, DMA time, transmission time on physical lines, *et cetera*) may become excessively high.

Note: The practical experiences in the design of switching systems has shown that small values of n (i.e., n equal to a few units) can still be pursued efficiently and with reasonable complexity. However, higher values of n lead to very complex and unstable behavior of common controls.

An approach that reduces the communication load among the units of a multiprocessor central control consists of the use of multiprocessing structures based on *common data memories* (see Figure 3.5). A common data memory is one viewed by each processor as if it were its own, similarly to its normal program and data memory. Via a common memory, programs executed by different CPUs may communicate among one another through common variables, as in the case of programs run by the same CPU. Being a common element of a highly reliable multiprocessor structure, common memories are duplicated. Their bottleneck factor lays in the fact that only one CPU at a time can access a common memory. If the CPUs are many and the number of accesses per unit of time is too high, each CPU must wait for each access, wasting its time until it can actually reach the desired variable. This causes situations similar to those already observed for multiprocessors communicating via their input output systems.

Note: To alleviate this problem, common data memories are organized in *banks* such that any two CPUs can simultaneously access different variables on different banks. This approach, if properly used, reduces the congestion of the CPUs around the common data memory. However, the problem remains and is only overcome at higher values of CPU accesses per unit of time.

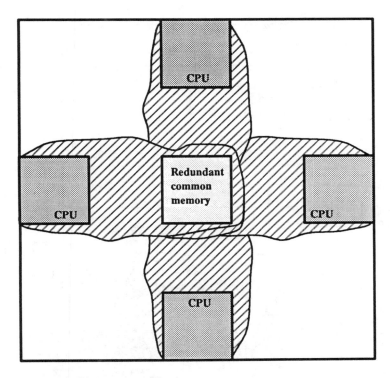

Figure 3.5 Central control with common memory.

By dividing the central control into a growing number n of CPUs, unless the assignment of programs is very carefully done, the number of accesses to common data memory by each CPU per unit of time grows as $n - 1$. Therefore, the total number of accesses to common memory from any processor grows, or at least tends to do so, with the square of n. Also, after carefully partitioning the programs, there are substantial difficulties in increasing n about 5–7 useful CPUs, plus others to be used as standby units in case of failure. This is what has resulted from practical experience with actual systems.

Thus far, the assumed partitioning strategy has been that of sharing programs among processors, with each processor executing its programs for the entire exchange. As an alternative approach, all software can be placed in each of the processors that execute it in a *load sharing mode*. An example is the GTD5 System from GTE (see Figure 3.6). In this case, several processors are connected to a common data memory. Some of them are used as *administrative processors* that together perform the functions of the service computer. The others are devoted to call processing. Call-processing CPUs have all the relevant application activities and share the processing load in a call-basis mode: calls are distributed by the

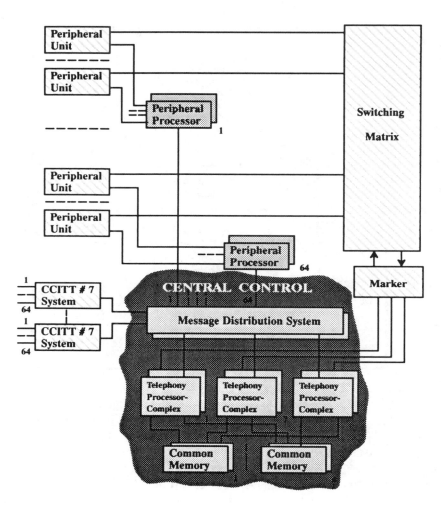

Figure 3.6 CPU load-sharing example (GTD5 system).

regional processors to those dealing with call processing in order to load them evenly. Inter-CPU communication occurs via both common data memory and a specialized device acting as a call dispatcher from regional to call-processing units. The GTD5 has been operated with traffic loads of several hundred thousands of busy-hour call attempts.

The potential criticality of the common control processing capabilities generally could be solved as a particular case of general-purpose multiprocessor architectures. For that purpose, one would need to invent a general multiprocessing structure with the following characteristics:

- It should be modular to cope with the wide ranges needed for switching exchanges;
- It should be highly reliable with built-in capabilities of automatic reconfiguration each time a failure occurs, or when a repaired unit is put in service;
- It should be provided with an operating system that makes it into a virtually unfailing processor, offering the multiprocessing environment required by the exchange application software;
- The same operating system should have a run-time support, which automatically allocates the processing activities to the available CPUs;
- It should be reasonably efficient in terms of the number of CPUs required to carry out the application activities of the exchanges, with reduced overload requirements due to the operating system.

So far, nobody seems to be close to achieving these objectives. The alternative approach consists of defining multiprocessor structures, specifically tailored to the nature of the application activities constituting the exchange software. The first examples of the chosen structures have already been discussed in the previous sections. Others follow.

3.5 Exploded Architecture for Common Controls: The Case of System 12

The switching system in which the fragmentation of the common control has been pursued to its fullest is System 12 from Alcatel (see Figure 3.7). A System 12 exchange is organized around a self-routing switching matrix, which is itself a network of elementary switching modules, all equal to one another. Each module is structured around a time-division bus connecting a module microprocessor and the access circuits of the PCM lines to be switched. The internal PCM lines (i.e., the lines connecting the switching modules and *peripheral modules*) include 30 + 2 channels. Channel 0 is used for synchronization, and channel 16 is for internal signaling among processors of contiguous switching modules (i.e., connected to each other with PCM lines). The frames of internal PCM lines have 16 bits each, instead of 8 as in normal PCM lines. This means that System 12 internal lines have a bit rate of $16 \times 32 \times 8000 = 4.096$ Mbit/s, twice as much as normal lines. In each frame of the useful channels (i.e., channels 1 to 15 and 17 to 31), the first 2 bits are used to specify the meaning of the remaining 14. This is done according to the following coding scheme:

00	Channel available,
01	Channel busy and forwarding a data call (*the other 14 bits are data*),
10	Channel busy and forwarding a speech call (*the other 14 bits are a voice sample*),
11	Channel busy but still in a marking phase.

An internal PCM channel can therefore be used to forward either a data

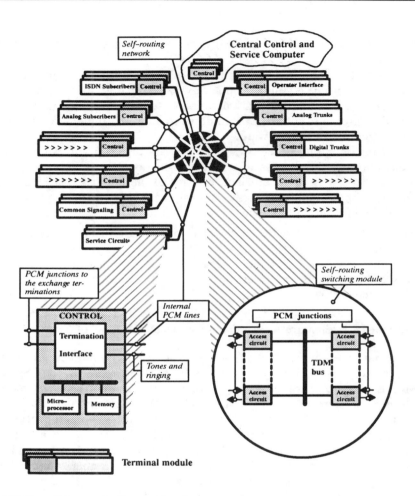

Figure 3.7 General layout of a System 12 exchange.

communication or a voice call. Moreover, during call setup, it is placed in a *marking* condition.

The switching matrix is connected to a set of peripheral modules of which there are two broad classes: those which consist of a processing unit only and of two internal PCM lines to connect it to the matrix; and those which have some kind of exchange termination unit, register, or exchange auxiliary unit around the processor. The peripheral units of the first kind (*processor-only*) are pieces of an exploded central control or of a service computer. As such, they can have attached to them computer peripherals such as disk, or printers. The peripheral units of the second class are regional processors of the unit to which they belong. Every peripheral unit has a standard interface consisting of two internal PCM lines with which the unit is connected to two different switching modules of the matrix.

On any channel of its two internal PCM lines, the processor of a peripheral module, irrespective of its class, may establish a bidirectional connection with the processor of any other peripheral unit. This connection can be used for either forwarding a call (which can carry either data or voice) or end-to-end data communication between calling and called processor. The first alternative can occur only between regional processors of line, register, and auxiliary units; the second also can occur in the case of processor-only peripheral units. Calling and called processors can, of course, also release a channel on which they are communicating. Call setup and release occur in a very few milliseconds. The switching matrix is self-routing in the sense that it does not have any centralized device to set up and release calls. These activities, in fact, are done autonomously, step by step, by each switching module through which every call is routed, based on the signaling information that all switching modules find, both on the useful channel from which the call is coming and on channel 16 of the related PCM line.

Let us consider the case of a hypothetical System 12 exchange with just two processor-only peripheral units: one providing the central control, the other performing the role of the service computer. Such a System 12 exchange would look like a regionalized switching exchange with the following features:

- The switching matrix, being self-routing, would not have a marker between it and central control;
- Signaling channels between central control and regional processors and between central control and service computer would be set through the switching matrix used for call setup. All this is done as necessary in each specific circumstance;
- To perform the circuit switching necessary in each call, the central control communicates to the peripheral module processor of the calling line the address of the peripheral module where the called line is located. Then the calling line processor starts the call setup through the matrix, which implements it step by step, without any intervention of central control.

The architecture of System 12 allows for the functions of both central control and service computer to be shared among any number of processor-only peripheral modules. Moreover, at least theoretically, these modules can be organized in either function-sharing or load-sharing mode. When using this feature, the signaling communication among the pieces of the common control (i.e., among regional processors, pieces of the central control, pieces of the service computer) is still done via the self-routing matrix, with each processor setting up and clearing signaling connections to the processors with which it needs to communicate, as dictated by each specific circumstance.

System 12 can exploit its built-in capabilities to divide the common control to obtain exchanges that are highly modular. However, the actual possibilities to divide the central control functions into too many processors remain affected by the considerations of the previous sections.

3.6 Modular Exchanges Implemented as Complete, Lossless Networks of Autonomous Switches

The practical realities of modern telephone exchanges have proved that the best approach is to keep all the application activities necessary in the exchange in the same central control; this is true with the possible exception of support and regional processors. However, such a solution conflicts with the need for modularity in telephone exchanges, and cost-effectiveness, independently of their actual dimensions. This is a particularly critical point because the size of an exchange can span over two or three orders of magnitude, from a few hundreds to over 100,000 terminations. This range of variation also requires similar flexibility in the processing capabilities required for the common control to execute its software activities. In this situation, in some way, to modularize the common control is advisable, while maintaining it adequately for the maximum dimensions foreseen for the exchanges. The criticality of these needs is further amplified by the fact that a continuously growing set of services is required in the exchanges, and the operation and maintenance features also tend to grow steadily. Both subscriber services and operation and maintenance features require processing time and, for the common control, constitute a substantial burden, which varies from site to site and from operating company to operating company.

The need to have powerful and modular common controls conflicts with the difficulties analyzed in the previous sections, and therefore we seek alternative architectures capable of maintaining most of the advantages of centralized common controls but without their intrinsic limitations and problems. On this subject, a possibility is suggested by choices being made ever since in the telecommunication networks where larger exchanges have always been implemented as a network of smaller switches.

Let us suppose (see Figure 3.8) for this purpose that we have available an *elementary telephone exchange* with a regionalized central control, which only requires the use of a support computer for operation and maintenance activities, not for call and signal processing.

Such an elementary telephone exchange will include, as well as its redundant central control (CC), a set of line units (LU), each provided with its own regional processor and specialized for the treatment of voice and signaling for a homogeneous set of exchange terminations (i.e., a set of subscriber lines, trunk lines, registers for common signaling channels, *et cetera*). In addition to the line units, the same exchange will have a *switching stage* (or matrix) connected to the central control through an *intelligent marker*. Nothing forbids adding to this already func-

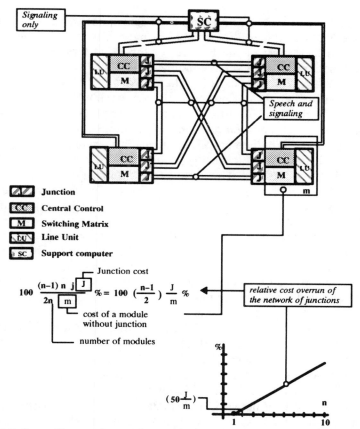

Figure 3.8 General layout of a complete mesh of autonomous exchanges.

tionally complete exchange a set of *PCM junctions* equal in number to a given parameter $K_{max} - 1$. Each of these junctions can be made equivalent to a multichannel PCM line without associated signaling. With the $K_{max} - 1$ PCM junctions, an identical number of signaling channels can be added, to be associated, on a one-to-one basis, to the PCM junctions. Each channel will terminate on its dedicated regional processor and, through it, will be connected to the central control of the exchange.

Elementary exchanges of this kind can be interconnected to compose networks of up to K_{max} nodes by PCM junctions and common signaling channels. The basic interest in such a structure is that, if all the nodes in the network are placed in the same site and synchronized to the same clock, the PCM junctions become very simple as they are reduced to pairs of cables of some sort, connected to the switching stages of two elementary exchanges. Thus, when the same clock is controlling all the nodes and these are placed in the same site, there is no need to implement full-fledged PCM lines using complete (and costly!) PCM terminations. In this situation, in fact, most of the features of the PCM terminations are not needed and the junctions become very economical.

Let C be the maximum number of simultaneous conversations which can be switched through the switching stage of each elementary telephone exchange. In this case, let C also be the number of PCM channels allocated in the same internal PCM junction. Thus, a call between any two terminations a and b belonging to two different nodes A and B in the network of elementary exchanges, once a route has been found for it in both the exchange A and the exchange B, can also be routed on a junction from A to B. This means that the mesh of junctions among the nodes is not only *complete,* but also *lossless.* This is because any call coming from an elementary exchange can be forwarded to another exchange, without the possibility of blockage due to the PCM junctions.

At this point, we should easily understand how the common signaling channel associated with each PCM junction is the common channel on which the central controls at either end exchange their signaling. Let us assume, as is the case, that the overcost D_S, due in each elementary telephone exchange to the presence of the junctions and related common signaling channels, is a marginal fraction of the total cost T_S. T_S is evaluated on the same switching exchange without junctions, common signaling channels, and the part of the switching stage due to the junctions. In this case, the overall cost of a network of K nodes ($K \leq K_{max}$) is very close to $K \times T_S$, which means approximately the cost of K bare nodes.

This circumstance remains true for every value of K_{max} that does not jeopardize the assumption that D_S/T_S is much lower than 1. Given this situation, for each $K \leq K_{max}$, the cost of a network of K nodes is equal, except for marginal contributions, to K times the cost T_S of a single elementary telephone exchange. Therefore, the entire exchange comprising a network of K nodes is intrinsically modular.

In a network of K elementary exchanges (see Figure 3.9) such as the one discussed, the worst case occurs when no call starts and terminates in the same module. Thus, every call is handled by two central controls: the one of the originating node A and the other of the terminating node B. Let now c_{it} be the number of call attempts for unit of time originated in each node. Altogether, the K nodes generate $c_{it} \times K$ incoming call attempts per unit of time. Given the assumption that each call attempt is handled by two nodes, the entire network of K nodes will have to treat $2 \times K \times c_{it}$ call attempts per units of time. This load, however, will be shared among K central controls. Therefore, each central control will have to deal with $2 \times K \times c_{it}/K = 2 \times c_{it}$ call attempts. This means that in the network considered here, each common control, for any $K \leq K_{max}$, will have to process, at the most, twice the number of incoming call attempts. Therefore, the network of elementary telephone exchanges would be capable of functioning correctly for

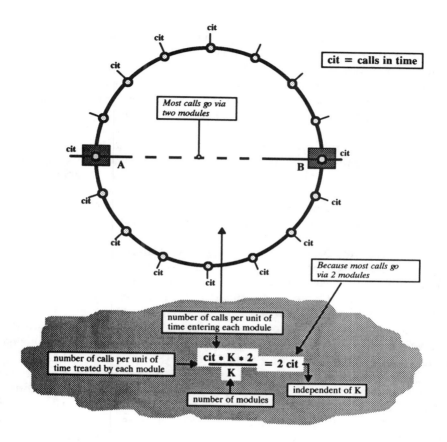

cit = calls in time

Most calls go via two modules

A   B

cit cit cit cit cit cit cit cit cit cit cit cit

Because most calls go via 2 modules

number of calls per unit of time entering each module

number of calls per unit of time treated by each module

$$\frac{cit \cdot K \cdot 2}{K} = 2\ cit$$

independent of K

number of modules

Figure 3.9 Average processing load in CC for a complete mesh of autonomous exchanges.

any $K \leq K_{max}$, provided that the processing capacity of each central control was equal to twice the capacity that would have been necessary if each elementary exchange were used as a stand-alone unit and not in a meshed environment. *This is the price that must be paid to implement the structure considered here.*

For this structure, the support computer is a centralized building block, which is necessary to provide every node in the exchange with unified operation and maintenance, as is also typically required for modular exchanges. The same support computer, however, can be connected via data links to more than one network at distant locations. This is done to implement unified management for clusters of exchanges, each typically remote from the others, and to reduce the cost due to the support computer on each subscriber line.

At this point, the question naturally arises as to how big an elementary exchange should be. The answer follows from our previous considerations: an elementary exchange should be as big as possible, up to the point at which its central control will be 50% loaded to have 50% still available for networking each elementary switch into a mesh of several switches. Industrial implementation based on this architecture (reference is specifically to the UT Line from Italtel) permit us to say that with a central control based on 16-bit CPUs and a processing capability around 0.7 MIPS, we can build an elementary telephone exchange capable of dealing with levels of telephone traffic above 7500 busy-hour call attempts. This allows elementary exchanges with up to 2000 subscriber lines or up to 350 trunk lines. The same implementations permit us to show that values of K_{max} around 16 are still possible and competitive; this permits implementing exchanges with capacities that can grow from a minimum configuration of one node to a maximum capacity of 16 times as much. This can be done without dealing with the complex and risky problems of partitioning the functions of a central control on more CPUs.

In the networks considered here, each central control is loaded with the same software. Moreover, each module knows the distribution of subscriber lines, trunk groups, and the remaining physical and logical resources throughout the exchange. This permits every central control to operate autonomously and to require the intervention of only the terminating node of each call. At a given moment, each central control does not know whether any trunk of the other nodes is free or busy. This fact may make use of the lines suboptimal. However, in most instances of practical interest, this limitation may be overcome by properly organizing the trunk groups and routing on each module.

The structure considered here shows some drawbacks when common signaling channels must be used; especially if signaling transfer point (STP) functions are also required. In this case, it is necessary to specialize one module where the STP activities are centralized. This need introduces a necessary exception to the simplifying rule that all nodes have the same software and run it as if they were autonomous exchanges. (This point is further investigated in Section 3.7)

3.7 Networks of Elementary Exchanges Connected by Structures Operating in Circuit and Message Mode

The intrinsic overcost D_S necessary in each elementary exchange to make it a suitable nodal element for the structure considered in the previous section is proportional to the number $K_{max} - 1$ of junctions and common signaling channels. Such an overcost, when summed for the entire network of elementary exchanges, increases quadratically with K_{max}, and therefore unavoidably diverges when the size of the desired macroexchange grows. Similar considerations stem from problems related to interconnecting the cables necessary for a complete network of K nodes. Such a network consists of $K \times (K - 1)/2$ connections, which must be built with $K - 1$ connections per module. Obviously, when K grows, things become complicated. Thus, the structure introduced in the previous section appears substantially advantageous because of its simplicity, for macroexchanges with relatively limited dimensions, but it is no longer competitive at higher dimensions.

In this situation, the analogy with telecommunication networks again suggests an appropriate adjustment of the same idea. The structure previously considered emulates the situation that would be found in a city of medium size, with an evolutionary environment in which the most advanced signaling systems can be used. In this case, in fact, typically desirable is to implement a network in which several terminal exchanges are interconnected with a complete mesh of lines and common signaling channels.

In situations where higher capacities are needed because of the size of the territory or the subscriber density, to allow such a meshed network grow endlessly is obviously not desirable. However, to implement a two-level network becomes preferable: a peripheral level and a transit level. At the first level there are still *terminal nodes*. However, instead of being meshed together, they are connected via one or more *transit exchanges*. Even in this case, of course, common channel signaling can be used and this facilitates introduction of a *signal transfer point,* connecting all the common signaling channels from the transit and terminal exchanges. Moreover, if the entire set of transit and terminal exchanges need to be operated globally, an *operation and maintenance center* can be organized and will be connected by data channels via the same signal transfer point to the transit and terminal nodes. With such a structure, the operator positions needed to operate and maintain the exchange will be connected to the central controls of the peripheral and transit nodes and to the operation and maintenance modules. Operator positions connected to the transit and terminal exchanges will have a logical link from there, via the signal transfer point, to the operation and maintenance modules. In this way, the workstations, irrespective of where they are physically attached, function as operator positions of the operation and maintenance center.

This general approach can be translated to the level of (see Figure 3.10) large exchanges. In this case, instead of terminal exchanges, we would be more correct to speak of elementary exchanges or, still better, *line groups*. The PCM lines will become simplified *junctions* synchronized by the same clock. The transit exchanges become *transit groups*. The elementary switching exchanges remain substantially identical to those in the previous section; moreover, they may be further simplified.

In fact, line groups need to have only two common signaling channels (for redundancy) to the signal transfer point, independently of the number of nodes in the macroexchange. At the same time, the line groups need only a number of junctions related to their outgoing or incoming traffic and made redundant for the usual reliability reasons. In any case, this number is largely independent of the number of nodes in the macroexchange.

The transit groups, if they are collocated with line groups and if they are controlled by the same clock, do not need to have PCM termination lines: in this

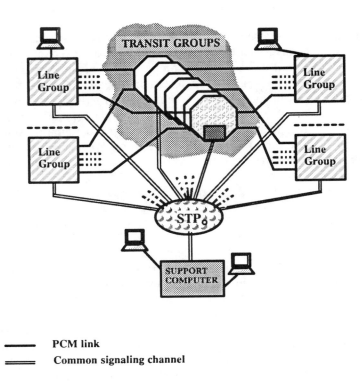

PCM link
Common signaling channel

Figure 3.10 General layout of a two-level network exchange.

case, they only make use of junctions. This substantially reduces their cost compared to that of the line groups. Each transit group, even if it does not typically operate at no loss, can be designed for greatly reduced blocking probabilities. Moreover, the possibility of alternative routing via other transit groups will further reduce the probability of a total blockage among line groups.

The considerations previously developed about doubling the processing capability, which is both sufficient and necessary in the common control of the elementary exchanges (or line groups) when used to implement larger exchanges, also apply in this case, but with some limitations.

Note: The communication process among modules in this case is more complex because the two terminal exchanges that deal with the same call attempt must also ask for the intervention of a transit group for it to make or suspend its connection through its PCM switching matrix. However, this necessity, when compared with the signaling and call processing by the two line groups, appears as a relatively limited processing inefficiency factor.

When the number of line groups connected to the same exchange increases, the potential bottlenecks in the architecture considered here may be the signal transfer point and support computer. This is because both of these elements, are centralized hubs in this architecture, which, because of their finite capacity, can jeopardize the growth of the exchange.

The signal transfer point, in general, can be perceived as structured around a *local area network* (LAN), to which a number of terminations are connected, equal to the number of common signaling channels between the same signal transfer point and the other modules of the exchange. The truly centralized resource in this structure is the bus (or the ring) that constitutes the local area network, which must be able to let all the signaling traffic flow among the exchange modules. As is known, technologies are already available that, based on *electrical buses* or *fiber optics,* allow us to predict practically limitless capabilities. Thus, we can state that signal transfer points, if properly designed, should not become actual bottlenecks in very large exchanges.

The limits of the capacities of support computers also can be overcome by making this unit modular. Support computers can thereby become networks of processors interconnected through either their local network or the message transfer point already assumed for the exchange. In a support computer, the cost effect of mass memories and data processing peripherals outweigh that of CPUs and core memory. This circumstance makes it important to implement the support computer by using powerful CPUs, which can therefore be few in number. We therefore conclude that multiprocessor support computers, even for very large exchanges, seldom need to have more than a few CPUs. An analysis of the organization of these CPUs in a multiprocessor environment is provided in Chapter 4.

By using structures conforming to the architecture analyzed here, we can implement exchanges with capacities adequate for above 100,000 exchange terminations. These exchanges may be either placed at the same site or distributed on a territorial basis. In this case, in fact, one can replace junctions with normal PCM links, where necessary, and provide the common signaling channels with the transmission support. Examples of exchanges architectures akin to the one discussed here are the ESS 5 System of AT&T and the large capacity exchanges of the UT Line from Italtel.

In our considerations thus far, we have assumed that the support computer is dedicated only to the implementation of operation and maintenance functions, excluding any functions directly related to call processings. However, we have also emphasized how the lack of a supervisory element to centralize the necessary coordination activities for routing and other aspects of call processing, for structures like the one considered here, leads to potentially suboptimal use of resources.

This circumstance suggests the opportunity to concentrate call processing functions in the support computer. These functions, by nature, are better centralized in a signal module. This choice has been effectively made by AT&T in its ESS 5. Of course, such a choice means that what has been described as a support computer becomes something different; namely, a module that also must be capable of treating calls. A support computer that treats calls must have redundant structures of the same kind as those adopted for central controls. Thus, for this kind of support computer, structures based on ordinary multiprocessors typically used in data processing are no longer possible. The inability to use relatively inexpensive and powerful technology counterbalances the advantages expected from putting call processing function in the support computer.

Note: As usual, these architectural comparisons are not the only ones to be considered when comparing systems. Other elements may overcome advantages or disadvantages based on just architectural considerations.

The line groups are, by definition, autonomous exchanges provided with junctions and signaling channels for communication among different macromodules; they are capable of interacting with each other according to a signaling language and design rules, which are uniform for the entire exchange. For simplicity, all these modules should be provided with the same software and they should offer to their terminations identical functions in each module. However, in a macromodular architecture this is not necessary. On the contrary, nothing forbids having functionally specialized macromodules. This means that, instead of putting the same software everywhere, some modules can be specialized for centralized functions that are not convenient to put in all modules.

As an example, we may consider the case of packet switching, as needed for ISDN. Packet calls, in fact, are only interesting to a limited number of subscribers and for a limited amount of traffic. If distributed on every macromodule, packet calls could cause inefficiencies in the exchange. However, one specialized macromodule, somewhat reduced in terms of terminal units, can be a better choice. The same considerations for the case of common signaling channels can be repeated here. One or a few specialized modules can treat the network-layer functions of common channel signaling and distribute messages to the user segments parts in all the exchange modules.

Thus, we emphasize how the macromodular structure, and that of System 12, are not so different from each other, but rather are two alternative ways of looking at such problems. When implemented, the two tend to converge toward rather similar solutions, at least, for the architecture of the common control and its software. The unifying factor comprises the problems related to partitioning the common control software, as discussed in Section 3.4. Partitioning is the most important limitation on the possibilities for specializing the macromodules in both the macromodular architecture and System 12. The same problem also shapes the call processing functions that can be effectively placed in the support computer, as done in the ESS 5 System.

CONCLUSIONS

Aside from the early solutions where the basic rationale was due to technological limitations that no longer apply, the architectures for digital exchanges that have been successful and competitive in the marketplace, may be approximately divided into three classes:

A.   Architectures with a regionalized central control,
B.   Architectures with an exploded central control,
C.   Architectures with a macromodular composition.

Exchanges of class A are structured as a bipole of a central control and switching matrix. Around this bipole there are a multiplicity of units, each provided with its own regional processors, where the processing activities related to the terminations are decentralized. Exchanges of class A, depending on the case, may or may not have a support computer.

In some cases (as in the case of GTD 5 System of GTE), the fragmentation of common control has been furthered by structuring it as a network of a few processors interconnected by a common memory. The significant example of architectures of class B is System 12 from ALCATEL.

Architectures of class C seek to implement larger exchanges as networks of smaller autonomous switching exchanges. Generally, class C exchanges can follow two alternatives; either they try to build a larger exchange by interconnecting a mesh of elementary exchanges through complete, lossless meshes, or they implement larger exchanges as two-level structures with line groups equivalent to terminal exchanges and transit groups, including circuit-switching nodes and message-transfer-point functions. In macromodular structures, the modules are interconnected by PCM junctions and signaling channels having a total cost that is typically marginal compared with that of the individual modules being used. As a rule, macromodular structures include a support processor to provide global management of the exchange. In some cases (for example, in the case of ESS 5 of AT&T), the support computer also performs some centralized functions for call processing.

Bibliographic Note

In this chapter the references have been intentionally limited to information on a minimal set of existing systems. Hereafter, a minimal list of references is given to provide the interested reader with a starting point for further studies on the problems considered in this chapter. For other switching systems not mentioned, the reader may refer to the literature.

REFERENCES

1.   Alcatel, *Revue des Telecommunications*, ITT, Vol. 56, No. 2/3, 1981. (Issue dedicated to the 1240 Switching System.)
2.   *AT&T Technical Journal*, July–August 1985, Vol. 64, No. 6, Pt. 2. (Issue dedicated to the 5ESS Switching System.)
3.   Ericsson, AX10 — System Survey, Ericsson Publication XF/YG 118429.

4.   A. Bovo and A. Bellman (Italtel), "UT100/60 and Electronic Digital Family of Exchange for Large Capacity Applications," *ISS 1984,* Florence, May 1984.

5.   S. Del Monte and V. Legnani (Italtel), "UT10/3, a Small-Medium Site Office for Local-Transit Application," *ISS 1984,* Florence, May 1984.

6.   M. Carsas and E. Pietralunga (GTE), GTD—5C: "A System for Local and Transit Applications: Evolutions from a Basic Architecture," *ISS 1984,* Florence, May 1984.

Chapter 4

Redundant Architectures and Software Quality

INTRODUCTION

In the previous chapters, we have shown how the use of one or more central controls plays a key role in most switching systems. We have also noted how exchanges with a classical structure have just one central control, whereas macro-modular exchanges employ a central control for each macromodule. The central controls of classical exchanges may be built around one or more *useful processors*. Central controls of macromodular exchanges are typically structured, at least in the most common systems, as pairs of devices, where only one unit is useful. The term "useful processor," explicitly used above, is meant to emphasize the situation, already briefly mentioned in the previous chapters, for which other processors are needed in a central control due to the reliability requirements of digital exchanges. The purpose of this chapter is to analyze terms, implications, and solutions associated with this situation which is central to understanding telecommunication switching systems.

For the moment, we will review the concept of common control as a monolithic entity, independently of its actual structure. If such a control fails, obviously, either the whole exchange or an entire macromodule will stop because, by definition, the commmon control includes all of the control functions of both a macromodule and the entire exchange.

Note: In the case in which support computers are also used and implemented separately from the central control, this critical situation applies only to the central control proper. However, an interruption in the operation of support computers just means a momentary inability to manage the exchange, without interruptions of telephone traffic flow.

Let G be the number of failures in the common control within a unit of time that, for the sake of clarity, is assumed equal to one year. Let L be the number

---

of lines controlled by the common control. Let R be the average time between a failure and its repair in the common control. R is to be measured in hours. The immediate result should be that the quantity $I = L \times G \times R$ is the average number of line-hours of outage per year in the entire exchange or in a macro-module. The parameter I gives a quantitative measure of the average damage associated with the tendency, which can be contained, but not eliminated, for any control to fail. In order to reduce I, we need to reduce one or more of its factors. This, however, cannot be done without incurring some penalties. As a starting point of this analysis, we can note how the reduction of L, in the case of the classical architecture, is simply untenable because the market demand for telecommunication networks rather forces L to grow. In the case of systems with macromodular architectures, the situation is a little less critical because the reduction of L in a macromodule inevitably means an increase in the number of macromodules per exchange. This factor cannot be taken lightly.

The containment of R is one of the basic requirements with which any switching system must comply. However, R results from the sum of three additional factors.

- The *delay* time necessary to the exchange to detect and signal a fault. This delay is very seldom longer than a few seconds;
- The *logistic* time necessary for the human technicians to initiate repair activities. This time can be confined under one hour only in exchanges that are 100% attended. Otherwise, the time becomes substantially longer because it must include the time for the technicians to go where the exchange is situated;
- The time necessary to *repair* the faults and to restore the exchange or macromodule to its normal operation. This time, except for completely anomalous situations, is less than a few tens of minutes (at least, such is the design prerequisite in the exchanges).

Because of these factors, R cannot be reduced to values smaller then one hour, even for large exchanges, where continuous (100%) attendance is possible. This minimum time grows to a few hours in the case of small exchanges and macromodules that are normally operated without around-the-clock attendance. In this situation, macromodules with significant numbers of lines, and even more so for entire exchanges, cannot permit unavailability of more than a few minutes per year. This fact requires values of G substantially below those of common processors and implies the necessity of circuit redundancy: controls that could have been implemented with only one processor will consist of two machines, one as a standby for the other; controls implemented as networks of a few useful processors will have to include further standby units, ready to operate automatically whenever a fault occurs.

The architectures for redundant centralized control will vary from system to system, and they form an original contribution of telephone exchange engineering to computer science. In the following sections, a general analysis is made of the solutions adopted in switching systems by organizing the many available options into three groups:

- *Tightly coupled microsynchronized central controls* (see Section 4.1);
- *Loosely coupled central controls* (see Sections 4.2 and 4.3);
- Central controls using several useful processors clustered around a *common memory* (see Section 4.4).

As in Chapter 3, we will avoid an analysis so deep that we may become lost in the details of each specific system. In Sections 4.1 to 4.4, we refer only to the architectures that are the most common to public switching systems. Our assumption is that, in a highly competitive environment such as that of telecommunication, a Darwinian mechanism of natural selection is in action, which brings success to only those solutions that best fit the actual needs of the networks.

Note: The considerations developed in Section 4.2 about a concurrent mechanism akin to the CCITT *specification description language* (SDL) as well as being useful for the topics of Sections 4.3 and 4.4, permit us to analyze some intrinsic characteristics of the exchange software. Therefore the considerations developed are used in the final sections of this chapter as a reference for the problems of software engineering in telephone exchanges.

The topics discussed in Sections 4.1, 4.3, and 4.4 specifically refer to choices encountered in central controls; excluding the case of regional processors or support computers. These other processors are considered in Section 4.5 (regional processors) and Section 4.6 (support computers).

Reliability and availability of a switching exchange depend not only on the architecture of its central control, but also on the characteristics of its software. This point is treated in Section 4.7. In the concluding section, we synthesize the topics developed in the chapter to provide a better understanding of the most relevant points.

4.1 Tightly Coupled, Microsynchronized, Central Controls

A first implementation philosophy for the central control starts from the idea of building commands as pairs of *twin processors* communicating via connection to their twin CPUs (see Figure 4.1). The two processors form a twin pair in the sense that, as well as having identical circuitry, they are loaded with the same software, which, in normal operating conditions, *they execute simultaneously and identically*. While executing each instruction, the twin CPUs communicate with each other about the details of what they are doing. This is to verify that they are indeed performing, at each individual step, the same activity. To operate in a microsynchronous mode, the twin processors are fed by the same clock signal, generated by a highly reliable device. For that purpose, a set of three clocks are typically

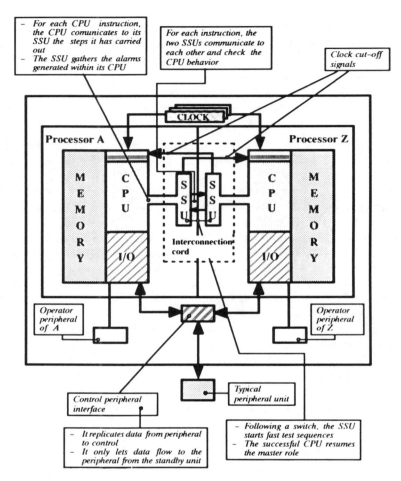

For each CPU instruction, the CPU comunicates to its SSU the steps it has carried out
The SSU gathers the alarms generated within its CPU

For each instruction, the two SSUs communicate to each other and check the CPU behavior

Clock cut-off signals

CLOCK

Processor A

Processor Z

M E M O R Y

C P U

S S U

S S U

C P U

M E M O R Y

I/O

Interconnection cord

I/O

Operator peripheral of A

Operator peripheral of Z

Control peripheral interface

Typical peripheral unit

- It replicates data from peripheral to control
- It only lets data flow to the peripheral from the standby unit

- Following a switch, the SSU starts fast test sequences
- The successful CPU resumes the master role

Figure 4.1 Microsynchronous double basic layout.

used, provided with a comparison circuit that chooses the proper clock in each situation in a majority logic mode. Microsynchronous operation allows for each twin CPU to decide, at each instruction, whether or not everything is working properly. Their analysis is correct as long as the twin CPUs do not make the same error at the same time. Such an assumption is realistic because the probability of two microsynchronized CPUs being simultaneously affected by the same fault can be made practically nil.

Each peripheral unit is connected by means of a specific input-output interface to both twins. All input signals from periphery to central control are *replicated* to both CPUs, so that each can see the data from their peripherals at the same time. In the outgoing direction, from central control to peripheral, each CPU behaves as if each peripheral were connected only to that processor. However, at every moment, just one element of the pair is characterized as *master*, with the other operating as *slave*. Only the output signals from the master CPU are actually forwarded by the input-output interface of the related peripheral unit. This means that, among other things, the interface units are capable of understanding, at each moment, which is the slave and which is the master. There are several easily understood mechanisms for such a function. The capability of switching between the two processors of a common control is continuously diagnosed on each peripheral interface. As a rule, an interface unit that is not fully operational is taken out of service, together with its peripheral.

The connection between the twin CPUs presents different characteristics from system to system. By accepting some freedom of interpretation, the choices can be modeled according to a functional design that considers the connection to be structured into two *synchronization and supervision units* (SSU), interconnected with one another and each to its CPU. At least at the conceptual level, each SSU is an extension of its CPU and continuously supervises the entire operation of its processor. For each instruction, the CPU executes an algorithmic sequence of microinstructions, characterized by arguments such as addresses of memory, labels of registers, data formatted as bit strings of variable length, operating codes to be executed by either the logic and arithmetic or input-output units, internal signals of the CPUs to command the transfer operations among registers or between registers and memory, *et cetera*. For each instruction to be executed, the CPU gives its SSU the corresponding sequence of microinstructions with their arguments. In this way, the SSU may summarize what is being done by its CPU in a sequence of *check points* to be sent to the twin SSU.

For efficiently detecting anomalous conditions, every unit of each processor is provided with its own checking circuits, which continuously verify *congruence conditions* of the machine. Typical congruence conditions are:

1. The verification that, among the bits of a given bus or register, a certain parity is mantained;
2. The verification that some specific events (typically identified by the edges of binary signals) are repeating according to defined timing rates.

The checking circuits memorize the results of their activities in specific registers accessible by the SSU, which uses them as complementary information to that which is received directly by the CPU. The two SSUs exchange the check points of their CPUs with each other and the contents of the registers giving the results from the test circuits. Every SSU, for each CPU instruction, checks the items sent to its counterpart against those received from it. As long as both SSUs find the comparison fully satisfactory, the two processors proceed with microsynchronous operation.

Obviously, microsynchronous operation, although it is the most frequent mode, does not continue forever. Figure 4.2 shows the evolution of operating

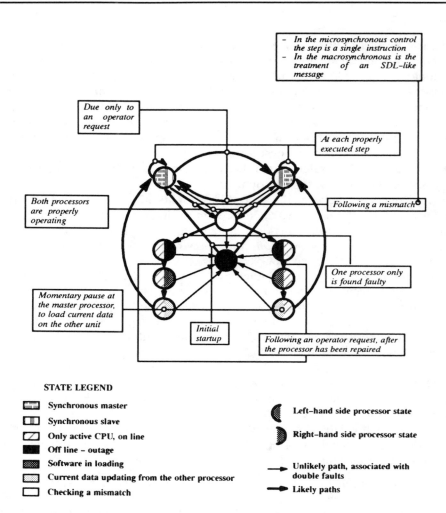

In the microsynchronous control the step is a single instruction

In the macrosynchronous is the treatment of an SDL–like message

Due only to an operator request

At each properly executed step

Both processors are properly operating

Following a mismatch

Momentary pause at the master processor, to load current data on the other unit

One processor only is found faulty

Initial startup

Following an operator request, after the processor has been repaired

STATE LEGEND

Synchronous master

Synchronous slave

Only active CPU, on line

Off line – outage

Software in loading

Current data updating from the other processor

Checking a mismatch

Left–hand side processor state

Right–hand side processor state

Unlikely path, associated with double faults

Likely paths

Figure 4.2 Evolution of the (main) operating states for a synchronous double (either micro or macro).

conditions for two processors that make up the control. The synchronous mode is interrupted each time either SSU detects a mismatch. When this happens, neither SSU can be certain which processor has failed or made an error. Therefore, the two SSUs must exchange a *synchronism outage* signal, which causes both to start a *self-diagnostic procedure*. At this point, the two processors suspend their current activities by maintaining their state and condition at the very moment of interruption.

Afterward the twins start an intensive self-testing program, identical for the two machines. Such a program is one of the critical elements of microsynchronous structures. It must, in fact, permit a quick, sure, and precise verification of the operating conditions of the entire processor, including its SSU. The program must be executed within a few milliseconds in order to suspend the current activities of the common control for a duration that does not disturb the exchange's application software.

For the self-diagnosis to be sure, precise, and quick, each twin processor is provided with a special set of instructions used to test the circuit functions of the device. Typically, such instructions lead to results detected either in the software of the same testing program or in the hardware by testing circuits similar to those already mentioned. These circuits, activated by the test programs, provide their responses on registers that are accessible by the SSU. The results of the test programs are detected by the SSU, which communicates them to its twin partner. Obviously, the SSU also is subject to failure. In this situation, the SSU must be exhaustively tested as part of the testing program. Typically, tests are implemented so that a faulty SSU provides its twin with either no indication (silence because of total failure), or clearly meaningless and inconsistent responses, which the twin can therefore interpret as a certain symptom of failure in its counterpart. According to this logic, each SSU is seen as a natural extension of its CPU so that a fault in the SSU is equivalent to an outage of the related processor.

The time interval within which the two processors must complete their self-test is a system constant, after which one of the following conditions occurs:

- *Single fault* — One SSU verifies that its own processor is properly working and the other is faulty; the other SSU does not disagree. (Note, however, that in practice it may.) In this case, the properly functioning processor assumes or resumes the master role and switches off the other to avoid any chance of disruption. The same processor restores the current activity of the exchange from instruction and conditions at which this activity had been suspended because of the mismatch.

- *Double fault* — Neither SSU can restore the functioning of its processor. This is a total outage due to a double failure, which, at least until the exchange is properly operated, should present an absolutely negligible probability of occurring.

- *No fault* — The two SSUs arrive at the conclusion that both processors are functioning correctly. Even this event is an anomalous situation, which exhibits (at least) one processor producing an error. The two SSUs will signal to their processors to resume their current activities, starting *from a common procedure,* which allows both to recover the errors that had caused the suspension of normal processing activities. In this way, the two processors return to the microsynchronous mode.

At least in principle, the no fault condition should occur only because of transient disturbances, although it may be due to an inability to detect every possible fault. These transient disturbances with the device properly operating, ought to be very rare. In any case, they are tracked by the central control and their repetition results in the execution of intensive maintenance activities under the supervision of operating personnel. In fact, to remove transient errors, one twin processor at a time must be taken off line and time-consuming testing procedures must be done to locate the transient faults.

For timely identification of faults, both permanent and transient, the twin processors include specific *cyclical test routines,* which keep their circuits under the control. These programs are executed concurrently with the rest of the software. The SSUs, however, are informed that the CPUs are executing cyclical tests instead of other software components (this can be done by means of suitable instructions, built into the machine language of the twin processors). In this way, the detection of a failure, whether transient or not, in one of the two processors results in a mismatch between the two CPUs, which the SSUs treat in a special manner. When an SSU detects a mismatch during the cyclical test, the unit does not suspend microsynchronous functioning because the test will switch off the faulty processor. However, the properly working CPU, once having completed the test, resumes its current activities. Then, when its SSU realizes that the other processor is out of service, it makes its own processor assume (or continue) the master role.

The verification of the proper behavior of the two processors must include all the interfaces with the peripheral units. These units, always must correctly react to the change in master role between the twin processors. This capability must be continuously verified by diagnostic programs within both processors (as usual, they are executed by the two processors when in a microsynchronous mode, but with the decisional capability reserved for the master CPU only). In principle, each unit that is not working perfectly is taken out of service by the central control. This requires, among other things, specific design rules in the interfaces between common control and peripherals.

Each time a processor assumes the role of master from another that had become faulty, the new master sends a signal to switch off the other unit (e.g., by cutting its clock source). A processor shut off can only be turned on following a specific command by a technician. This avoids, for certain kinds of errors, one machine in a master mode being forced into a slave mode by its counterpart before the latter has been properly repaired.

When, for any reason, the microsynchronous mode is suspended, except for catastrophic situations, either processor can control the exchange while the other is being repaired. In this situation, the two processors work independently of each other: the on-line unit runs the exchange software (including its cyclical tests to avoid a fault in the on-line processor confusing the entire exchange); under the control of operations personnel, the other executes the repair procedures. To make this possible, each processor is provided with a specific console used to perform the relevant repair activities when necessary.

After a processor has been repaired and suitably tested, it must be put on line. To do so, a copy of the exchange software must be loaded onto the processor, identical to what is operational at that moment on the other processor. This activity can be done either via the twin connection with a transfer from the other processor, from the support computer, or from a mass memory. The load via the twin connection is executed in a DMA mode, with the twin processor dedicating one machine cycle every N, where N is large enough not to disturb the current activities of the exchange software and small enough to load the other processor in a reasonable time.

With the programs loaded, the processor must be put back on line with all the current data of the twin CPU; both the data in its memory and whatever is stored in its registers. This activity requires the intervention of the SSU. Typically, there will be a momentary suspension of the on-line CPU to avoid changes in its current data while a copy is brought to the CPU that is still off line. There will be a transfer, via the connection between the two processors, resulting in the full alignment of the twins. At this point, microsynchronous operation can be resumed, starting from the instruction at which the still solely on-line processor had been suspended to update its twin, still in restoral phase.

The total integrity of programs and data and their perfect identity in the twin processors are prerequisites for a properly operating microsynchronous central control. Therefore, the loading of programs and alignment of data are done with procedures that verify the correctness of the transfers. Different procedures are used for this in various switching systems. As an example, for the case of programs and data memory, we refer to *parity control* mechanisms on words and bytes transmitted and to the *verification* of starting or ending addresses in the memory spaces used for the allocation of programs and data. For the content of registers, however, the techniques used are those of reading the registers after having written them to check that what has been written is exactly correct.

A microsynchronous common control is a rather complex and intrinsically costly entity. This is due, first, to the presence of the twin interconnection, which is atypical of processors in common use for data processing applications. Then, there is the necessity within the common control to implement cyclical check procedures and self-diagnosis programs. For such programs to be feasible, the circuitry of the entire control must be designed with the relevant testing circuit and the SSU to be capable of diagnosis given the tremendous quantity of errors that are reasonably possible. The design of a processor by identifying, *a priori*, all the ways in which it can become faulty, is still more complex in cases such as those discussed here where speed and precision are particularly critical. However, once all these problems are correctly solved, a microsynchronous pair of twins becomes a control, which, independently of the nature and structure of its software, provides a switchover between master and slave processors from the very instruction at which the master unit fails. This ability makes the microsynchronous twins among the most effective redundant structures for applications requiring very high reliability.

4.2 An SDL Architecture for the Software of Switching Exchanges

In the considerations developed in Chapter 2, we emphasized how the software that must operate within a telephone exchange is structured into a multiplicity of elementary activities, all of which are executed concurrently and in real-time. To synthesize the situation, we can state that in a common control the situation is as if a multiplicity of *active entities* were operating in it, working concurrently, interchanging solicitations, and operating on common variables. This circumstance, as indicated in Section 4.3, is the conceptual basis for a second family of redundant central controls. Hence, this section, in which the goal is to provide a possible formulation of the intrinsic nature of telephone software.

The suggested formulation is based on software architecture closely akin to the CCITT specification description language. The considerations developed here should help in better understanding the typical characteristics of telephone software and central controls of loosely coupled macrosynchronous pairs. We should, however, stress that the software architecture considered here would be but one possible example for the architecture of telephone software, stated only as a descriptive tool. Each switching system, in fact, has its own software structure, which differs from case to case, even if all of them are consistent with the intrinsic nature of the telephone software.

The software structure can be introduced by a succession of *rules* (see Figure 4.3):

Rule 1: All of the software run by a common control, consists of the concurrent interaction of a multiplicity of *interrupt drivers* and *waiting state monoprocessors.* The interrupt drivers are related to the interrupt signals generated in the hardware of each processor.

To understand the following rules more easily, interrupt drivers and monoprocessors can be perceived as *individuals,* each capable of performing its own activities. Both interrupt drivers and monoprocessors may be dormant. They remain so until they are *triggered.* Interrupt drivers are triggered by interrupt signals, and monoprocessors by solicitations. When triggered, interrupt drivers or monoprocessors perform a sequence of actions, after which again becoming dormant.

Rule 2: Each interrupt driver is a program started by the occurrence of the interrupt signal handled by that driver. Once started, the driver acts until its natural completion, which is the execution of a return statement to the program interrupted by the signal.

Rule 3: Priorities may be set for the interrupt drivers. The priorities adopted will be consistent with those of the hardware interrupt leads in the processors. The

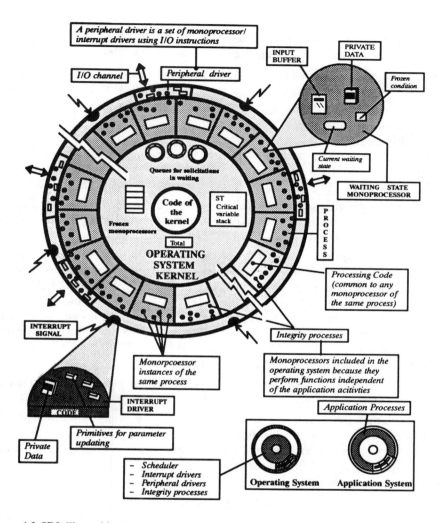

Figure 4.3 SDL-like architecture.

return statement of each driver *G* results in the resumption of the previously interrupted entity, which can be:

1. Another interrupt driver,
2. A monoprocessor,
3. The *message waiting cycle* (see rule 34).

Rule 4: A monoprocessor is an entity capable of either (a) staying dormant in one of its *waiting states,* or (b) being active, executing an algorithmic sequence of *processing actions.*

Rule 5: Each monoprocessor m is characterized by its own set of N_m waiting states, distinct from each other and identified by the positive integers from 1 to N_m.

Rule 6: The monoprocessors (a) exchange, (b) send to the output peripherals via specific drivers, and (c) receive from the interrupt drivers *solicitations,* characterized by:

1. An *operating code,* which identifies their type;
2. Possible arguments;
3. An originating entity (which can be either a monoprocessor or peripheral driver);
4. The urgency level requested for the execution of that solicitation.

Rule 7: A *driver program of a peripheral unit* is itself a set of one or more monoprocessors and interrupt drivers, code of which includes the machine instructions necessary for the physical operation of the relevant peripheral.

Rule 8: Once in a waiting state s, a monoprocessor m remains there until it receives a solicitation specifically directed to it. When this happens, the monoprocessor m executes a multiple selection group, such as, for example, a PL/1 SELECT or a PASCAL CASE, with its alternative paths labeled with the operating codes of the messages. In the selection group, a specific path is provided for each solicitation expected by the monoprocessor m while waiting in s. To these alternative paths, an escape clause is added (such as the OTHERWISE in PL/1 and ELSE in PASCAL) for all messages considered unacceptable when m is in s.

Rule 9: While a monoprocessor is active, it is deaf to the solicitation that it may receive, and remains so until it arrives in another waiting state.

Rule 10: Each monoprocessor m is provided with an input buffer, having a capacity of only one solicitation; in this buffer, the other entities put messages to m and will do so only if the buffer is empty. Each monoprocessor remains in a waiting state as long as its input buffer remains empty. When leaving a waiting state, a monoprocessor empties its buffer.

Rule 11: Each alternative path (including those of the escape clauses) is a *BEGIN . . . END* block, which describes the actions by the monoprocessor in that case. This block uses the primitives *NEXT, SEND, CHANGE, TIMEOUT,* and *AL-TIME,* discussed in the following rules.

Rule 12: The primitive NEXT (index s of an arriving waiting state), means that:

- the monoprocessor must be put in a waiting state s;
- the control of the program execution must be returned to either the next monoprocessor (see rule 28) or, if there are no waiting solicitations, to a *cycle for waiting messages* (see rule 34).

Rule 13: The primitive SEND (name st of the structure where a complete solicitation has been stored; it coordinates the monoprocessor to which the solicitation must be sent; priority is assigned to it), specifies a request to send the solicitation written in the structure *st* to the monoprocessor and with the priority specified in the other two arguments of the primitive. The result of a SEND may be either positive or not. The response is positive if the input buffer of the destination monoprocessor is found empty by the SEND. When it does or does not succeed, the SEND also sets a binary variable *SENT* at 1 or 0.

Rule 14: After each SEND, a block *IF SENT THEN . . . ELSE . . .* must be set up to specify actions to be taken by the monoprocessor.

Rule 15: Also, the interrupt drivers may send solicitations to the monoprocessors by using the SEND with all its associated arguments.

Rule 16: The interrupt drivers do not use NEXT and messages cannot be received by the drivers because they are not provided with a state structure.

Rule 17: In addition to the input buffers of the monoprocessors, each interrupt driver operates on its own *private variables,* accessible only to it.

Rule 18: Each interrupt driver is designed with a set of subroutines to update the driver's *private parameters*. These subroutines are the only mechanism available for the other monoprocessors to operate on the interrupt drivers. (*Note*: Such a choice is due to the need to avoid monoprocessors and drivers conflicting with each other in the use or modification of configuration parameters of the interrupt drivers).

Rule 19: More monoprocessors $M_1 . . . M_n$ belong to the same family if and only if their codes are identical, except for an (implicit or explicit) *instance index,* which differs in value for each monoprocessor of the family and can be used as a variable, an index, or a pointer.

Rule 20: A *waiting state process* is the code, with its parameters set by an *instance variable* for all the monoprocessors of the same family. The waiting state process reflects the behavior of the monoprocessors in the same family.

Rule 21: The *address* of a monoprocessor is an ordered pair ($p;i$) in which p specifies the process and i the instance to which that monoprocessor must be referred within the family.

Rule 22: The *critical variables* of the processes are those which (a) some processors modify and others read; (b) cannot be modified independently of the state of the monoprocessors related to them.

Rule 23: With respect to the critical variables, the states of a monoprocessor are divided into two classes:

- the *terminal* states: when the monoprocessor is in these states, any critical variables can be updated without disturbing that monoprocessor;
- the *transit* states: these are defined in terms of logic complementarity of the terminal states.

Rule 24: The modification of any set of critical variables by a monoprocessor must be executed by means of the following primitive:

CHANGE (critical variable = new value, . . . , critical variable = new value)
WAITING $(p' - i', \ldots, p'' - i'')$

This includes the variables for modification and their new value; it specifies also the monoprocessors (p,i), which the system must await to come to a terminal state before the actual modification of the critical variables can be done.

Rule 25: Within the same processor, one CHANGE can be executed at a time, without any overlapping between consecutive CHANGES. The execution of a CHANGE is therefore refused by the system when another CHANGE, already accepted, is waiting to be completed. The acceptance of a CHANGE is indicated by a system variable called *ACCEPTED*.

Rule 26: Each CHANGE must always be immediately followed by a statement of the kind:

IF ACCEPTED THEN NEXT (. . .) ELSE . . .

in which the (alternative) actions that must follow the CHANGE are specified.

Rule 27: After an accepted CHANGE, the monoprocessor that generated it, because of the NEXT following the ACCEPTED, enters a waiting state, where it remains deaf to any other input messages until the CHANGE has been completely finished.

Rule 28: In each processor, only one monoprocessor at a time is executed. Each monoprocessor is started when it has to carry out a solicitation in its input buffer and maintains control until it arrives at a NEXT. When more monoprocessors' input buffers are full, that which has the solicitation with the highest priority is executed. When there are solicitations of equal priority, the longest waiting solicitation is executed first. In any case, no processor that is blocked according to rules 30 and 31 is started.

Rule 29: When a CHANGE is accepted, it implies the following immediate effects:

- Preparation on a system buffer T_S containing the addresses of the variables to be updated together with their new values;
- Activation of *freezing traps* on the monoprocessors, specified after the WAITING clause and found in a transient state by the CHANGE;
- *Blocking* of the monoprocessor specified after the waiting clause and found in a terminal state by the CHANGE;
- Load in the variable *total* of the number of monoprocessors specified after the WAITING clause and found in a transient state by the CHANGE.

Rule 30: The NEXT included in an accepted CHANGE, as well as executing the functions specified in the previous paragraphs, *blocks* the monoprocessor to which it belongs; in such a condition, this monoprocessor cannot accept other messages as input until released according to the following rule.

Rule 31: The NEXT, as well as its other functions already observed, only for the case in which its argument is a terminal state, does the following:

- If the freezing trap of the monoprocessor is active (see rule 29), it decreases by 1 the variable *total* and blocks the same monoprocessor, which remains insensitive to other messages directed to it until released according to the following point of this rule.
- If after having decreased *total* by 1, this variable becomes 0, and therefore all the monoprocessors implied in the last accepted CHANGE have arrived at a terminal state, it executes the transfers specified in T_S, releases all the monoprocessors blocked, and sets the conditions of acceptability for a further CHANGE primitive.

Rule 32: The *TIMEOUT* (time to be waited; label) primitive, activates a clock having the symbolic name *label*. Unless previously suspended with an *ALTIME* primitive, as soon as the time specified in *time to be waited* has elapsed, it generates a solicitation into the input buffer of the monoprocessor that had invoked it. This solicitation contains the value *label* as an argument. If the input buffer is already busy, the time-out message is discarded without any possibility of retrieval. Time-out messages correctly stored in the buffer are treated like those generated by the SEND.

Rule 33: The ALTIME (label) primitive determines the suspension of the clock symbolically named *label,* previously activated by the same monoprocessor. If such a clock does not exist, or has already been suspended, ALTIME has no further effect.

Rule 34: In the initial starting conditions of the application software, all the input buffers of the monoprocessors are empty. Therefore, no processor can be activated. Thus, until a new solicitation arrives, the *cycle in waiting* will be executed (see rule 12). Solicitations may be generated following the hardware interrupts treated by the drivers. Typically, some of these signals are used together with the relevant drivers and a set of monoprocessors to implement the functions of starting, resuming, and suspending all of the software of the control element. In each circumstance, the cycle in waiting, as soon as it finds a message to an unblocked monoprocessor, passes control to it.

Rule 35: To know under which circumstances to send each solicitation, each monoprocessor and interrupt driver, in read-only mode, has access to the waiting states where any other monoprocessor is placed.

In addition to the monoprocessors, an implementation of an architecture characterized by means of the previous rules must have an *operating system kernel* including the following functions:

1. The cycle in waiting;
2. Execution of the primitives NEXT, SEND, CHANGE, TIMEOUT, and ALTIME;
3. All of the hardware interrupt drivers;
4. Time-out management.

From such a kernel, to complete what is commonly considered as an operating system of a central control, we need to add monoprocessors that implement the following functions:

5. Drivers for the peripherals;
6. Supervision of the processor's actual behavior;
7. Diagnostic and switchover for redundant processors;
8. Exception handling procedures (see following paragraph).

Note: The exception conditions mentioned in rule 8 must be used to treat *erroneous* messages, which should not have arrived at that monoprocessor in such a state. This typically means that the group *BEGIN . . . END* associated with these conditions is not expected to change substantially from case to case and is to be structured into standard procedures for error recovery, consisting of algorithmic sequences of subroutines included in a *library* of the operating system. By doing so, we can introduce systematic mechanisms into the software for tracking residual erroneous conditions to be used during both the program's debugging phase and the initial *seasoning* period (see Section 4.7) when a new software release is brought into service.

All Other Monoprocessors in the Application System

In the software architecture suggested here, the monoprocessors are the elementary components with which to build the exchange software. The many activities, segments, *et cetera*, in which this software is structured are assembled into organic sets of monoprocessors.

To make the communication mechanisms among monoprocessors sufficiently efficient in terms of speed and CPU processing time, they must interact, not only by exchanging solicitations, but also operating via a *common database,* in which each writes what itself and the other will need to read. For this purpose, a distinction should be made between *private variables* and *common variables.* A private variable is internal to just one monoprocessor, which is its only user. The initial value when the monoprocessor is activated does not affect the result of the related processing. Private variables do not belong to the common database.

Each monoprocessor can also contain some *internal permanent variables,* which it may update each time it is invoked. These variables and the current waiting state make up the memory of what the monoprocessor had done up to that moment. *The permanent internal variables are a section of the common database.* A monoprocessor has access, but in read mode only, to the other monoprocessor's, internal and permanent variables. Similarly, it can read the current waiting state of the other monoprocessors.

There are, finally, variables that belong to the common database, and can be modified and read by several monoprocessors: these are the *critical variables* stated in rule 23. They must be treated only by the primitive CHANGE to avoid the possibility of incongruities among monoprocessors.

The concept of *terminal state* corresponds to the circumstance that sees the monoprocessors more stable in some states than in others. Thus, for example, a monoprocessor that describes the behavior of a telephone line will have some terminal states which state that the line is out of service, inactive, broken, blocked, *et cetera*: these are its terminal states. In addition, other *transient states* will specify the operating conditions of the line during the signaling phases of a call. As we know intuitively, with a monoprocessor in its terminal state, the critical variables can be changed without risk. Conversely, when in a transient state, the change of these variables is potentially dangerous.

Equally meaningful, from a practical viewpoint, is the concept of *sequence* as a path of a monoprocessor between two terminal states. In reference to the previous example, a sequence becomes each of the possible types of calls that can occur on a line.

In many situations, the exchange software carries out concurrent activities characterized by a different degree of urgency. In these circumstances, the use of parameters to state the urgency level for each SEND allows organizing on different priority levels the activation of monoprocessors related to activities that, during overload conditions, may be postponed until the processor's load again becomes acceptable.

The choice of canceling time-out messages for which the input buffer of the destination monoprocessor is already busy is related to the circumstance for which the time outs are typically started to verify that given messages arrive within a certain time; typically, if the input buffer is busy, the message should already have arrived and the time out has not occurred.

Situations where the time-out counters are used to measure the time may be implemented with waiting states in which a counter is started and the time is measured through messages from time outs that are the only solicitations to be awaited. In this situation, in fact, as soon as the time has elapsed, the time out will reach its destination monoprocessor. More complex situations can be implemented by using monoprocessors specifically dedicated to the treatment of time outs.

4.3 Loosely Coupled Macrosynchronous Controls

In industrial logic, the most critical aspect of microsynchronous doubles is in the uniqueness of the circuit solutions to which they bring, especially with reference to the need for interconnections with the characteristics discussed in Section 4.1. This uniqueness results in serious difficulties in the use of circuit components commonly available on the market and suitable to the needs of common processors, rather than those of microsynchronous controls. The difficulties of employing components for which the enormous market volumes of data processing applications grant cost-performance levels that cannot otherwise be attained imply further cost overhead for microsynchronous controls that are not marginal. This situation has created an interest in other architectures capable of maintaining as much as possible the advantages of microsynchronous doubles, but at the same time reducing their unique hardware.

Approaches in this direction can be classified under the common architecture of doubles, which, for reasons that will become clear later in this section, may be defined as *macrosynchronous* or *loosely coupled*. A macrosynchronous control (see Figure 4.4) maintains a structure with two processors, which, under normal conditions, operate in parallel.

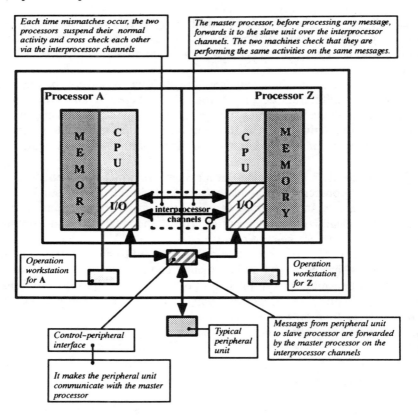

Figure 4.4 Macrosynchronous control.

In this case, however, the interconnection is replaced by a pair of input-output channels, typically implemented by two pairs of parallel ports. Each peripheral unit is also connected to the control via an interface that links it to both processors. However, the communication process between peripheral and control occurs only with the master computer. As will be shown later, the slave processor receives messages from the peripherals via the master CPU. The two processors of a macrosynchronous structure are each provided with a clock and are loaded with substantially equal software, but with some distinction in the operating system.

In macrosynchronous controls, the characteristics of the telephone software are exploited; namely, its structure into monoprocessors. On this subject, the architecture adopted for the exchange software becomes a central factor. This architecture varies from system to system. In this situation, for the sake of explanation, reference is hereafter made to the software structure considered in the previous section. The considerations developed, except for some details, are substantially valid for the arbitrary choice made here. The software of a common control is broadly structured into two parts: *operating* and the *application* systems. Specifically, the operating system includes the following functions:

1. Solicitation dispatcher to the monoprocessors and activation of the monoprocessors;
2. Execution of the SDL primitives invoked in the monoprocessors and interrupt drivers (NEXT, SEND, CHANGE *et cetera*);
3. Drivers for the interrupts, peripherals and channels between the two processors;
4. Management of the macrosynchronous integrity, including all related aspects, as discussed in the following sections.

Note: However, the monoprocessors that carry all of the telephone traffic and the operation and maintenance of the exchange devices external to the central control make up the application system.

The operating system makes the central control appear to the application system as if it were a single processor without an interprocessor communication channel, or the need to have two processors acting more or less at the same time on every peripheral. This virtual processor is capable of concurrently carrying out the application activities and it never fails, at least as long as the exchange is properly operated. The application system is identical in the two parallel processors. However, the operating system offers different variants, which are executed respectively for the case in which one processor is master, slave, out of line, *et cetera*. This software package, including all its variants, is loaded in both parallel processors. However, each of them will execute at any given moment only the variant related to its current operating state.

The communication between central control and peripheral interfaces is organized into messages, similarly to the case of monoprocessors. In this situation, the scheduler, included as a part of the master's operating system, interleaves requests of messages in input from the peripherals with the execution of internal messages already put in queue by the monoprocessors and the interrupt handlers. The master's operating system executes one message at a time, which may either be internal or have arrived from outside the central control. The master's operating system starts the relevant monoprocessor, which remains active until it finishes its activity by means of a NEXT. The master's operating system gives each received message to the slave unit, from both outside and inside, before sending the message to the relevant monoprocessor for execution. Afterward, the master waits for a positive answer from the slave processor. Only after it has received such an answer, the master sends the next message to the monoprocessor for the appropriate processing.

The slave processes every external message that it gets from the master, as if received directly by the peripheral unit. Conversely, for each internal message (which means each message generated by a monoprocessor or an interrupt handler), it verifies that this message is the one that it would have executed due to the current situation of its message queues. Only in such a case does the slave send to the master processor a positive acknowledgement and process the message as if it had been generated by the slave.

The mechanism of mutual locking on the internal messages permits the two parallel processors not only to verify that everything is working properly, but to remain aligned with each other. This is because the faster of the two will be stopped by the slower one until it arrives at the same result on the same message. Only the master unit of the central control sends output messages to the peripherals. However, the messages from the slave processor are blocked by the peripheral interface.

The synchronization procedures between the two parallel processors are executed on the two connecting channels by a protocol, which allows each machine to know whether the other is operating properly. As usual, the principle applies that meaningless answers and lengthy silences make certain that the other processor has failed. The use of two macrosynchronization channels is necessary for reliability reasons because the interruption of the line between the two machines eliminates any possibility of parallel operation. Also, there are numerous situations of uncertainty in the interaction between the two processors on a channel that can be efficiently overcome only by again trying something that has been found to be wrong on the other channel.

As for the case of microsynchronous doubles, the parallel processors of a macrosynchronous central control are also symmetrical and do not have an element that operates to resolve conflicting functional conditions. However, in this case, there are two processors provided with their own software, which are connected via two common input-output channels. This connection allows the two processors to start, in the case of a conflict, a cross-checking program, where each assumes itself to work correctly and tests the other. This approach is meaningful given the assumption (which is fairly certain in the case of large integrated circuit technology) according to which a processor, when it fails, either stays silent or communicates nonsense. Thus, when a parallel processor fails, the other can (in all statistically relevant cases) correctly identify such an event and assume (if necessary) the master role in resuming its activities from the last message being processed when the fault became known.

Once the two parallel processors have detected a mismatch (see Figure 4.2), the two operating systems suspend the current messages and execute test programs, which lead to one of the following results:

1. *One of the processors is put out of service and the other assumes the current activities starting from the last messages not yet completely processed.* This circumstance is possible because the slave processor, under normal conditions, executes the same processing functions as the master unit and has its data constantly updated so that it can immediately take over in case of failure.

2. *Both processors are found to work correctly.* This circumstance should never happen because, without failures, no mismatch should occur. However, when this situation happens, the two processors have insufficient information to decide which one went wrong. This fact implies the necessity for the operating systems of both machines to execute a resumption procedure equivalent to the release of every call in the signaling (transient) phase with most of the calls already in conversation left undisturbed.

3. *Both machines are found faulty.* This situation must be very unlikely to occur because it brings about a complete blackout, which can be recovered only through a total system resumption.

Each time a switchover occurs in the master-slave roles between the two parallel processors, the possibility exists that one input message already sent from a peripheral unit to the master has not yet reached the slave. Because of this possibility, the interfaces between common control and peripheral must be capable of accepting the retransmission of messages already forwarded from the peripheral units.

Similar uncertainty exists for messages sent from master processor to peripheral units just before the role switchover between the two parallel processors. To overcome this difficulty, the operating system of the slave must store each message that it would have sent to the peripheral unit until the former unit starts processing a new input message from the peripherals. In this manner, when a switchover occurs, the operating system of the new master processor first sends the output messages accumulated during the processing of the last input messages that it had received while still operating in the slave mode. This behavior implies the possibility for the peripheral units to receive the same message twice. (This is nontrivial as peripherals must be able to decide whether to act on a repeated message.) Once more, this implies the need for interfaces and peripherals to have some processing capability. Moreover, between peripheral units and common control a communication protocol is needed to cancel duplicated messages and to retransmit unreceived information.

In a macrosynchronous system, there are infrequent chances of loss of synchronism where neither processor is continuously faulty. In these cases, the two parallel processors must be provided with a suitable resumption procedure, which maintains most of the active conversations undisturbed and releases, as necessary, those calls still in a signaling phase. This capability cannot be located in the operating system because it is strongly dependent on the monoprocessors that make up the application system. Therefore, this procedure must be implemented as a set of processes as part of the application system. When the need arises, the operating system of the master processor activates the resumption procedure with an internal message to the initial monoprocessor. In such a case, this internal message is also passed to the slave processor, which is thus forced to execute the resumption procedure.

As for microsynchronous systems, the two parallel processors of macrosynchronous controls also may start working independently of each other. When this happens, (see Figure 4.2), one unit executes the exchange functions, and the other operates off line. To put the second unit again on line, it must be loaded with the same software as the first on-line unit. This can be done either by using a mass memory connected to the common control, or by copying, via the parallel channels between the two processors, the on-line unit's software. In the latter case, the master processor operates in an alternating cycle mode (which means that it uses one every N machine cycles to make the transfer). The frequency of the cycles being chosen is to minimize the disturbance of the current exchange activities, while keeping as low as possible the total time necessary for the off-line machine loading.

In the case of macrosynchronous controls, the two parallel processors also must be maintained under continuous scrutiny to ensure that they are working properly. Typically, they include some self-testing hardware to permit each CPU to detect failures such as parity bit errors in accessing the memory or faults in the internal busses. These events cause hardware interrupts that result in the processor suspending itself. Within the operating system of the parallel processors, a consistent set of test programs must be included to be cyclically executed by the CPU, one piece after the other, during and between the treatment of useful messages. Negative responses during the execution of these activities cause the machine in which they occur to stop itself.

Within the operating system, programs must be provided to test the interaction among the parallel processors. These programs consist in the exchange of diagnostic messages between the two machines, for each of which a well defined response is expected. Messages and responses are sent on the two parallel channels connecting the processors. Each time a processor does not receive the expected responses from its counterpart, one unit stops the other by interrupting its clock and, if necessary, switches over as master.

A macrosynchronous double is a kind of control that is relatively easy to implement by using electronic components commonly available on the market. However, most of the specific features implemented by special hardware in microsynchronous doubles are realized by software. This circumstance causes macrosynchronous commands to consume substantial amounts of their CPU time to carry out operating system functions. Thus, less CPU time remains available to deal with the application system activities. Therefore, a careful definition of all the functions related to macrosynchronous operation is the critical aspect for this class of central control architectures.

Macrosynchronous controls do not allow for a switchover at the very instruction during which the failure occurs. This requires them to have intelligent interfaces with the peripheral units and communication protocols to the CPUs. Moreover, the likelihood of misalignments between the two parallel processors, although reduced, is not nil. This implies that there will remain recovery activities during which at least the calls in the signaling phase will be perceptibly disturbed for the subscribers.

4.4   Central Control with Common Memory

In the two previously considered architectures, both CPUs of the central control must execute the same useful activities. This is required to allow the slave unit to keep in its memory a fully updated version of all current data as necessary to replace the master CPU when it fails from the point at which it has failed. This is done because each processor in the central control has its own memory. Also, architectures based on *common memories* have been successfully tested in practical switching systems. A central control architecture based on a common memory, at first glance (see Figure 4.5), can be seen as a macrosynchronous control in which the parallel processors have a substantial part of their memory in common. As a centralized component, the common memory must be duplicated. To enhance further the system reliability, the common memory can be structured into disjoint *banks*. Each bank covers a segment of the memory's physical address and consists of two *planes,* one redundant with the other. A fault on a bank does not affect either plane of any other bank. The banks offer a complete, nonoverlapping coverage of the common memory. To write a variable in common memory means that the CPU will update the variable in both planes. The reading of a variable is normally executed in but one plane; the second plane is used only if anomalous circumstances arise. This is possible because common memories are provided with parity checks that allow deciding whether data have been read correctly.

A common memory is seen by an accessing processor as its normal working memory with (almost) the same efficiency and speed. However, *common memory*

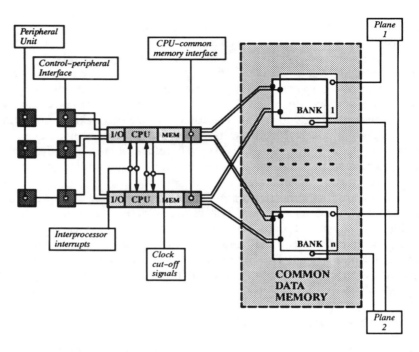

Figure 4.5 1 + 1 central control with a common memory.

is used only to store data, not programs. The connection between each CPU and the common data memory is via a specialized interface that typically generates to its CPU an alarm interruption each time it fails to access an operation. A failure on this interface causes a global fault in its CPU.

The availability of the common data memory allows reducing the channels between the two CPUs of a macrosynchronous control to two logical interruption wires: one for each direction. (One wire for each direction can be used instead of two because the reliability of the circuitry associated with these wires is very high as compared to that of the CPU to which they are connected.) By exchanging signals on the two interruption wires and data via common memory, the control processors may interact with each other as in the case of macrosynchronous controls. This ability allows introduction of routines into the operating system that makes each processor perform self-tests and others for the processors to test each other. Also, in this case, we can assume that at most one CPU is faulty at any time and the one that remains operational can detect failure in the other. When this happens, each CPU is provided with outgoing wires used to switch off (typically by cutting off the clock signal) a faulty processor. This capability ensures that no further harm is caused to the central control.

In a common-memory-based control, a CPU operates as the *principal unit,* the other as a *standby unit.* Under normal conditions, only the principal CPU executes the application programs. Conversely the standby unit interacts with the other, and its operating system runs test routines. In this way, the standby CPU can quickly replace the principal one as soon as any failure is detected.

Note: In any circumstance, the application system is executed by only one processor. Each peripheral physically connected to both processors interacts at any moment with only the CPU executing the application activities.

Also, in the case of common-memory-based controls, there must be tools for a timely and precise identification of faults in any CPU. This need is fulfilled by use of a combination of checking circuits and diagnostic programs as part of the operating system. The testing circuits generate alarms that are sent out from one processor to the other as interruption signals. The diagnostic programs are run by both principal and standby unit, and completely cover faults that cannot be done by the hardware circuitry without adding excessive costs to the processors.

The functioning of a common-memory-based control may be easily described by assuming that the exchange software is structured according to the reference model in Section 4.2. This software is based on an operating system implementing a virtual device, which (see Figure 4.6):

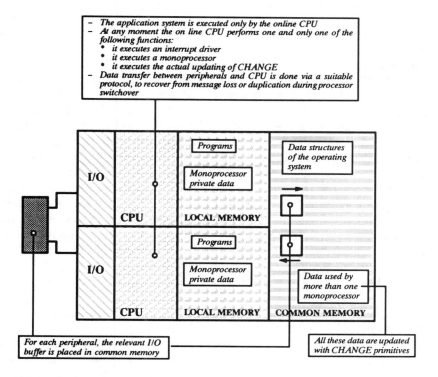

Figure 4.6 Application software organization in a 1 + 1 central control with common memory.

1. Makes the monoprocessors and the input drivers evolve according to their rules;
2. Implements the self-test and cross-checking functions between principal and standby unit;
3. executes the switchover functions;
4. executes the input-output functions with the peripheral units.

Also, in this class of architectures, the communication processes between common control and peripheral interfaces must be done according to message-oriented protocols that allow recovering from message duplication and loss during processor switchover.

The data transfer between control and peripherals is only done by the processor that at that moment is performing the *on-line* functions while the other is *in waiting*. (Normally, the on-line and the principal processor are the same unit; however, the opposite is true when the principal CPU is out of service.) Messages are transferred from peripheral interfaces to a special buffer in the common memory. One buffer is provided for each input port. Similarly, one buffer is available in the common memory for each output port to transfer the messages directed to it. These buffers also contain the necessary information for a CPU to decide the stage at which a message transfer has arrived at that moment. This allows (following a switchover) the incoming on-line machine to resume the correct input-output activities.

All the input buffers to the monoprocessors are placed in the common memory together with any other variable used by the operating system to perform its run-time activities. In the on-line processor, the operating system executes one monoprocessor at a time. Any variable that is not totally *local* to a monoprocessor and has initial or final values meaningful within or outside that monoprocessor is placed in the common memory and can be updated only by means of a CHANGE primitive. The primitive is executed by the on-line processor only after the actual termination of its invoking monoprocessor. However, at invocation time, a CHANGE only memorizes the address and new value of each concerned variable.

Each time a switchover must be performed, the choices stated here allow the newly on-line processor to start its application activities without loss of continuity: with the other processor executing a monoprocessor, the same monoprocessor is repeated; with a CHANGE being performed, the same is repeated. Any SEND primitive, when repeated, results only in the overwriting of the same message on the same buffer. This is also true for output messages to external peripheral units.

In the case of common-memory-based controls, the software in the processors must be identical. This software, however, will be the sum of the programs that each processor may need to run in any circumstance. In a given situation, each processor will only run the variant pertaining to its operating condition. With one off-line processor that needs to be brought back in service, the first step is to load it with the run-time software. This can be done from either an available mass memory or the on-line processor. In this second case, the common data memory may be used as a transfer channel operated at a speed that will not disturb the ordinary processing activities of the on-line processor. Once the transfer is completed, there is no need to update the data because they are already available in the common memory. The off-line processor can therefore be returned to service in a *waiting* condition. After being stabilized in this state, the operator may eventually command a switchover between the control processors (see Figure 4.7).

In common-memory-based controls, the need arises to align the planes of a common memory bank. This is typically done by the on-line processor, which, while performing its normal activities copies in background, one plane onto the other, one memory position after another. At the same time, during this copying

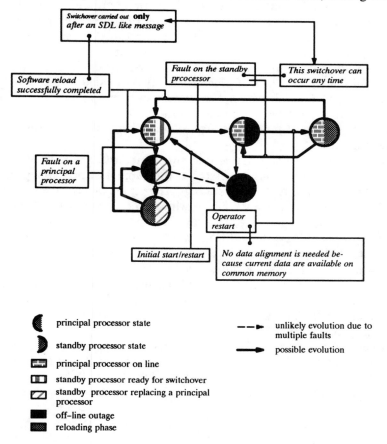

Figure 4.7 Evolution of the operational states in a multiprocessor central control with common memory.

activity, the on-line processor writes the variables in both planes. By so doing, at the end of the plane loading, the two instances of the bank have identical values.

Common-memory-based controls allow saving CPU time as compared to macrosynchronous architectures. This is because common memories do not need to waste time aligning the parallel processors with each other at each message. This advantage, however, must be traded with the costs of a common data memory that cannot be dismissed as negligible.

Far greater interest in common-memory-based architectures is due to the fact that they permit generalization of the case of $1 + 1$ structures to that of $N + 1$ processors, where N units are principal and one is standby. In $N + 1$ structures, the N principal processors share the load of the application system, while the standby unit is there to replace any principal processor that may fail. Also, the peripheral units of the central control are shared among the principal processors as each of them is connected to the standby and to *one* principal unit. Moreover, the same connections as in the case of a $1 + 1$ structure are replicated between each principal and the standby processor (see Figure 4.8).

The same software with a minimum of modifications in the operating system due to the multiplicity of principal processors may be loaded in the $N + 1$ units of the control. Configuration parameters written in common memory define the monoprocessors that must be run on each principal processor. The standby unit may replace any processor put out of service either because of failures or due to operator commands.

An $N + 1$ structure is a device that simultaneously executes N monoprocessors at a time: as many monoprocessors as the number of principal processors. This circumstance implies that the operating system must include the capability of avoiding any chance of concurrent conflict, which, in the case of $1 + 1$ structures, are not possible because they execute only one monoprocessor at a time. This means, among other things, that a SEND must not only check that the receiving buffer is empty, but also seize it before using it and see that no other processor is trying to do the same at the same time. Similar considerations apply for the other SDL primitives (e.g., CHANGE) and the dispatching mechanisms used for solicitations and monoprocessors. On this consideration, several approaches are possible, depending on the machine language, processor hardware, and ingenuity of the operating system designers.

As a general design rule, the N processors are expected to operate at full load so that the standby unit can replace only one principal unit at a time. This means that the number N can be as large as the value N_{max} such that the probability is still negligible of having two of the $N_{max} + 1$ CPU fail within the average time needed to repair the one that has failed first. In cases where higher values of N are necessary, the principal units may be divided into subgroups of N_i principal processors each. Each group will have its own standby processor. All of them will have the same common data memory. By doing so, the $N + 1$ structure is gen-

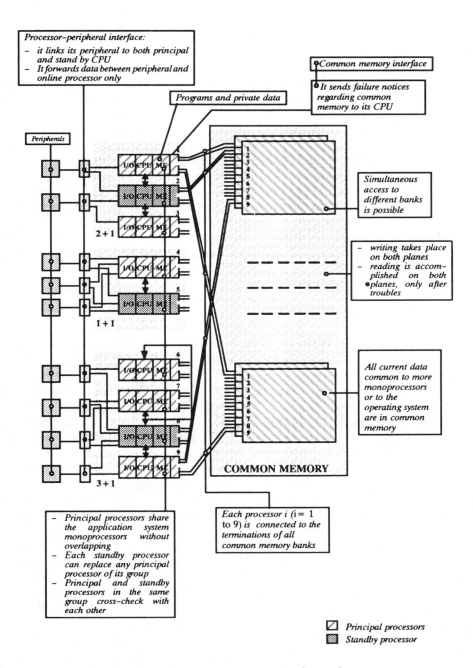

Processor–peripheral interface:
- *it links its peripheral to both principal and stand by CPU*
- *It forwards data between peripheral and online processor only*

Common memory interface

It sends failure notices regarding common memory to its CPU

Programs and private data

Peripherals

2 + 1

1 + 1

3 + 1

COMMON MEMORY

Simultaneous access to different banks is possible

- *writing takes place on both planes*
- *reading is accomplished on both planes, only after troubles*

All current data common to more monoprocessors or to the operating system are in common memory

- *Principal processors share the application system monoprocessors without overlapping*
- *Each standby processor can replace any principal processor of its group*
- *Principal and standby processors in the same group cross-check with each other*

Each processor i (i = 1 to 9) is connected to the terminations of all common memory banks

Principal processors

Standby processor

Figure 4.8 Structure for a multiprocessor common memory central control.

eralized into a multiple $N_i + 1$ to which the same considerations apply, except for minor adaptations.

To limit the congestion of the common data memory due to the fact that only one processor at a time may have access, the common data memory banks are structured as independent units such that any two processors can access any two banks at the same time.

Note: Every principal processor is assigned one or more banks, which are used to make the principal unit's monoprocessors communicate with each other. No other principal processor accesses the banks assigned to the others to avoid unnecessary access conflicts in common memory. Monoprocessors located in different principal processors must communicate via banks common to their principal units. To avoid congestion due to this circumstance, a careful partitioning is necessary for the monoprocessors of the principal processors: any two monoprocessors should be placed in different principal units only if they do not communicate extensively. The practice of switching systems has proved that this objective is achievable for only small values of N.

4.5 Regional Processors

Thus far, we have systematically referred to the case of central controls, ignoring that of regional processors and similar units. This section is intended to fill that gap.

In most cases, regional processors (see Figure 4.9) do not need to be made redundant because a fault in any of them implies only the outage of exchange units that may be provided with an on-line spare to be used whenever a failure occurs. As an example, let us consider the case of registers. These units are engineered on an exchange to have more units for each class than necessary to cope with ordinary peak traffic conditions. In this way, when a register fails, the remaining ones can still manage the exchange activities.

The case of units that are not interchangeable, for example, subscriber line units, may be treated in two ways: either these units are made small enough to reduce both the likelihood and effects of their outages; or their regional processors are made redundant. In this case, however, the needed redundancy schemes are much simpler than those of central controls. This is because of the existence of central controls that view each individual instance of a regional processor as a different peripheral unit: a different unit for each duplicate of any redundant regional processor. For each instance of a regional processor, the central control uses a driver through which it supervises the unit to see whether it is operating properly. When a failure occurs, the central control can switch to a standby instance after having updated it with the last value of any configuration parameter stored in the failed regional processor. The actual problems that arise in each specific

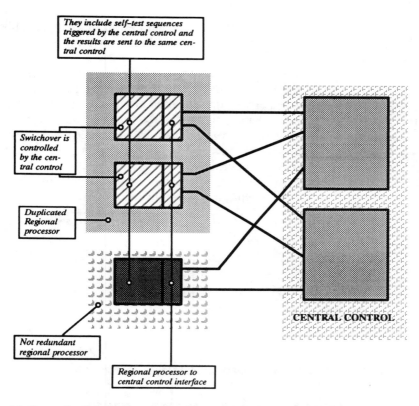

They include self-test sequences triggered by the central control and the results are sent to the same central control

Switchover is controlled by the central control

Duplicated Regional processor

Not redundant regional processor

Regional processor to central control interface

CENTRAL CONTROL

Figure 4.9 Connecting links between regional processors and central control.

circumstance vary from case to case. We can observe, however, that, in general, the software of the regional processors is SDL-like and can therefore be modeled according to the rules and definition of Section 4.2. This generalization allows us to understand how the problems related to fault detection, resumption of operation, and reconfiguration can be solved at the architectural level of the SDL structure for each regional processor.

4.6 Service Computers

Because of the functions that a service computer performs, it may go out of service for short times without affecting the traffic carried by the exchange. However, any chance of incorrect actions that could jeopardize the current exchange activities must be avoided. Neither can a service computer be allowed to become unavailable too frequently or for long periods of time. The latter circumstance, considering the logistical time needed for fault repair and recovery, also leads to redundant structures for these machines. The architectures that best fit this case are substantially akin to those of ordinary computer clusters based on local area networks (see Figure 4.10).

LAN structures allow for modular service computers, which can be tailored and made to grow according to the size of the controlled exchanges. The functions to be performed may be shared among the clustered processors according to strategies including reconfiguration capabilities that take into account the resources actually available at any moment. Critical functions that need to be provided following a processor failure may be assigned to two processors: one executing as first choice, the other as a backup alternative.

In the case of service computers, the capability must be provided to detect the occurrence of faults. This is typically accomplished by a set of test routines

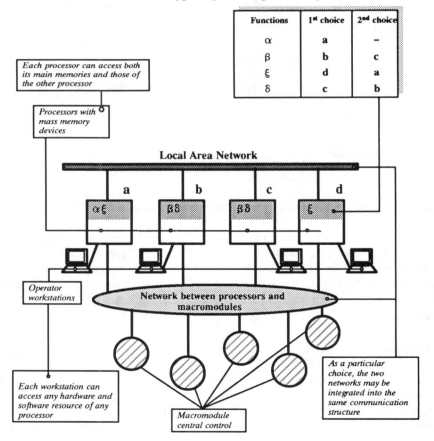

Figure 4.10 Interconnection layout of the service computer.

that allows the processors in the cluster to check one another. These routines are added to the operating system of the cluster and are executed in background, via time-sharing with the application system activities. Warnings and alarms generated by the test routines are shown on specific workstations and permit the operators to know when they must start reconfiguration activities on the service computer.

Reconfiguration procedures result in complex sequences of programs, which are part of the cluster operating system. They may be executed by momentarily putting the service computer off line from the exchange central control.

The interactions between central control and service computer tend to be organized as an ordered sequence of *transactions*. Each transaction is well defined in scope and duration. During its execution, the two machines send information to each other, but do not update their data. Data are updated only at the end when both central processor and service computer realize that the relevant transaction has been successful. While executing each transaction, the machines involved check the consistency of the data exchange. If any inconsistency arises, the central control communicates this fact to the service computer (possibly by a line connected to a different cluster processor from the one where the inconsistency has been detected) for its checking routines to decide what action to take. Inconsistencies should be detected by the service computer or central control, and should result only in losses or duplications of messages recovered by the communication protocol between the two machines.

In general, the entire set of application activities done by the service computer is a flow of transactions structured according to the architecture embedded in the processor's cluster operating system. Each transaction produces, in addition to temporary data that vanish with it, *permanent* results of one or more of the following classes:

- Information sent to the central control;
- Information sent to remote computers;
- Information displayed at operator workstations;
- Data stored in permanent (disk) files.

When a service computer is reconfigured, most of the transactions being run are aborted and, eventually, each resumes from the beginning. Therefore, as already noted, the central processor must be provided with suitable mechanisms to deal correctly with information from the service computer. Similar considerations apply to any other remote computing system. However, the data shown at operator workstations pose only minor problems due to the human capabilities to recover easily from corrupted information.

Much more difficult problems relate to the management of permanent files, some of which contain sensitive information such as billing data and the database describing exchange structures and configurations. These files must be replicated on different physical media. Moreover, the service computer software must be capable of correctly updating all copies, detecting anomalous situations, and aligning the copies with each other when the need arises. These features must be included as part of the service computer's operating system so that the application activities refer to virtually incorruptible databases, where they can read and write without any concern about duplications.

4.7 Software Quality

In the previous sections, problems related to the exchange's reliability had been considered with the assumption, more or less explicit, that the cause of the problems was hardware failures. This idea was dominant until the end of the 1970s. However, the experience gathered from the field has shown that the actual situation is more complex. In fact, hardware failures are, on the average, responsible for only 25% of maintenance activities, with the remaining 75% due to procedural errors made by people (35%) and residual errors in the exchange software (40%). Furthermore, the technological developments of the 1980s in the field of integrated circuits have made available components that are both more powerful in terms of processing capability and more reliable. As a consequence, modern exchanges can be designed with a very high complexity in their hardware, where the likelihood of most possible faults is hardly detectible. This situation justifies simplifying assumptions (such as those accepted in the previous sections) about the hardware failures to be considered in the maintenance software and further reduces the percentage of switching outages due to hardware problems.

The effect of human factors as causes of outages demands a great deal of care in the design of person-machine interfaces. They must be both friendly to the operators and meticulous in checking their actions at the level of every step in the person-machine interaction. This demands a lot of software, and still more ingenuity in designing it. Nonetheless, the total levels of protection that can be achieved are not absolute because the operation and maintenance procedures also depend on the actions performed by the operators; namely, the introduction of correct data on the machine and the correct and timely replacement of faulty units. As an example, we mention that the overall reliability levels of a common control are provided only if, after a failure has been detected, the proper actions are taken within very stringent amounts of time.

The complexity levels in software exchanges are so high that it is practically impossible to remove 100% of any design and implementation errors. On this topic, the experience of the previous decades has shown that no universal panacea is available to produce totally error-free software at reasonable costs and in reasonable amounts of time, as needed in highly competitive market environments. Neither is any such panacea likely to be discovered in the next decade.

The complexity of the exchange software is due primarily to its dimensions, which can be expressed, with everything included, in order of magnitude of above one million lines of code. A small part of this (about 25%) is used for telephone traffic processing. The other is part of the operating system (about 20%) and the operation and maintenance software (55%). The number of simultaneous transactions that the exchange software must be able to run is strongly related to the number of calls the exchange must be able to handle. Figure 4.11 shows some reference parameters. All the telephone traffic must be executed in real-time environments that require response times by the software of a few tens of milliseconds for calls in signaling phase, fractions of seconds for calls in conversation, and a few seconds for operation and maintenance activities.

Apart from the *normal conditions* usually encountered in the network lines, the exchange software must be able to deal with unpredictable situations, which may be due to any signaling error and remote exchange faults. Also, the operation and maintenance software must face any hardware failure, be able to recover from any residual software error, and must correctly treat any human action related to the exchange.

The design and implementation of such software is among the most difficult industrial software activities, and the removal of any possible error is practically impossible due to the inability to simulate every conceivable circumstance that an exchange may face, leading to faults in the hardware, errors on the lines, or inadequate human actions. The best possible approach in this respect consists of proving the software against a reasonable set of statistically frequent cases. Then the software is put in the field, in a normal environment, and made to work while repairing the residual errors still present in it.

Multiprocessing requirements (for a local exchange with about 10,000 subscribers

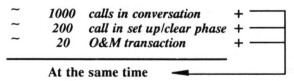

~	*1000*	*calls in conversation*	+
~	*200*	*call in set up/clear phase*	+
~	*20*	*O&M transaction*	+

At the same time

Maximum response time of the software (real–time requirements)

☐	*For signaling activities*	*10– 100 ms*
☐	*Call processing activities*	*100–1000 ms*
☐	*Person–machine*	*1 – 3 s*
☐	*O&M transactions*	*1 –10 s*

Figure 4.11 Real time requirements for the exchange software.

To make this possible, the exchange software must be designed to accommodate the unpredictable in a systematic and planned way. This can be done by disseminating in the monoprocessors, in the appropriate place, *exception conditions* (see Section 4.2, rule 8) or OTHERWISE clauses that trap any eventual inconsistency. When an OTHERWISE clause is activated, the events will be fully tracked in a systematic way to permit accurate understanding of what has happened and to produce a software modification that takes into account the circumstance not yet properly handled. Following any unpredictable situation, the software cannot crash, but must be able to overcome it, possibly losing calls still in the signaling phase, seldom losing any call already in conversation, still less frequently via an automatic resumption of the entire exchange, and with or without reloading the software. In this respect, the *quality* of a software release is measured by considering the traffic carried by the (software) exchange *versus* the average losses of traffic due to the errors made on calls. Under actual circumstances, by measuring the performance data in the field and tolerating the residual errors, still to be removed, that occasionally cause trouble. Figure 4.12 shows some reference figures that are reasonable requirements for an item of software to be considered of acceptable quality.

To mitigate the complexities of the exchange software, the choices related to the following topics have a particular effect on the achievement of satisfactory levels of quality at reasonable costs:

1. The consistency of the software architecture;
2. The programming languages used;
3. The organization of the development teams;
4. The software production logistics.

These points are therefore briefly treated in the following paragraphs.

A software architecture consists of a comprehensive set of rules adopted for the organization of the entire exchange software into *modules* or *components*. These components are either *database modules* or *program modules*. They must be organized into two nonoverlapping sets, the *operating system* and the *application system*. The operating systems include all the software necessary for the concurrent

❏	**Exchange down time**	**2h/40 years**
❏	**Rate of calls improperly handled**	**~ 10^{-4}**
❏	**Rate of undercharged calls**	**~ 10^{-5}**
❏	**Rate of overcharged calls**	**~ 10^{-6}**

Figure 4.12 Requirements for reasonable software reliability.

execution of multiple programs, interactions with the processor peripherals, and management of the processor resources. In the case of redundant controls, the operating system also includes the software that manages the diagnosis and integrity of the entire control to make it appear to the application system as a never-failing machine, as long as standby processors are available to replace faulty units.

The operating system must also include all the *software primitives* invoked and used by the application system activities to communicate with each other, via both solicitations and common data. In this manner, the operating system implements the architecture of the application system. This architecture includes a comprehensive definition of the way in which the application system must be decomposed into elementary modules and how these modules can be made to interact. This should be done by a set of well stated rules, should be easy to grasp for the programmers, fully controlled by the operating system, and efficient for the design of the application system. A (partial) example of such an architecture is the monoprocessor structure introduced in Section 4.2.

As already mentioned, the application system architecture must include the programming rules used for the systematic treatment of exceptional conditions. The actions following an unexpected event must include a minimum set of functions by the operating system, carried out independently of the software written by the application system programmers. These interventions must include generalized tracking of anomalous events and subsequent resumption of the application system. Further specific, *ad hoc,* initiatives to be decided on a case-by-case basis can be added as part of the application system.

The use of high level programming languages, sufficiently efficient from the viewpoint of the CPU time that they consume, can greatly increase the quality of the exchange software, while reducing its cost/performance ratio. This fact is especially evident in the case of languages provided with compilers characterized by extensive and precise compilation-time error-checking capabilities. The use of languages based on *typed variables* and allowing for both compilation-time and run-time control on variables and data is particularly significant because the data used in the exchange software are largely non-numerical in nature and use bit strings, the configuration of which are coded to represent values of abstract variables defined on a case-by-case basis. These variables need to be defined by their type, including the set of values permissible for each. This is done to permit a timely verification that no illegal value is being stored in any variable, which would bring catastrophic corruption to the software loaded in the exchange.

What is actually needed for an efficient design of the exchange software is the use of languages that permit defining variables by *types* and data by *classes*. Types and classes, as termed here, are equivalent to algebraic entities defined by their values and the rules to process them. These algebraic entities will be the data stored in the variables handled by the software. The programming languages must allow for broad definitions of types and classes and for a pervasive check, at both compilation time and run time, that no inconsistency arises in processing variables and data with respect to their formal definitions. As run-time checking is rather consuming of CPU time, it should be performed selectively, where and when it is actually needed. The definition of types and classes needs to be feasible in terms that describe the software data in an easily readable and straightforward manner to the programmers. Only by being able to understand easily the meaning and role of any variable in the software, can a programmer be efficient in performing his or her activities, finding remaining faults, and fixing them without introducing new problems.

Less significant, especially for the case of central controls and regional processors, is the use of *structured programming* techniques. The exchange software is naturally expected to be organized into small pieces, each one dedicated to the processing of a specific message. Because of their reduced dimensions, these segments are easy to read, even if not written in a structured way. Again, what is mostly needed is not a formal structuring of the processing statements, but that of the data on which these statements operate. The data, especially handled by many monoprocessors, need to be well structured to reduce the risks of errors and inconsistencies.

The software of a switching system, because of its size, complexity, and life span, results from the organized activity of large staffs of people over long periods of time. This circumstance emphasizes that what is critical for the quality of the software is the organizational environments in which the developments occur. These environments must include the implementation of planning strategies that are comprehensive for all phases of the software development processes: from the early design to the final maintenance. A technical discipline has been born from the analysis of problems and solutions of these strategies: namely, that of software engineering.

In addition to the points stated in the previous paragraphs, the exchange software does not present any special peculiarity demanding its own software engineering approach. Therefore, methods and planning tools adequate for normal process control software also apply to the case of software for switching exchanges. Generally, the following points appear of major concern:

- The totality of the software package, with all its related documentation of any kind, from design, to implementation, to user manuals, must be conceived according to a well defined set of *documents,* each defined by its scope, content, and meaning; each is constantly and consistently kept up to date during the life cycle of the software.
- For each document, the organization in charge of writing, updating, checking, and supervising it must be clearly identified; each organization bears clear-cut and strongly enforced responsibilities.
- Revisions in the design, development, and maintenance of the exchange software should be regularly scheduled, and anticipated as much as possible against the final delivery of each software release.

Unlike the case of hardware, for which the engineering disciplines are naturally enforced by the ongoing dialogue between design (i.e., research) and production (i.e., factory) people, in the case of the software, all activities tend to be handled by the same sort of people, whose organization is more a matter of goodwill and choice than a need due to the existence of people with different cultures, motivating factors, and behavior. Combined with the fact that the overheads related to the enforcement of software engineering practices are substantial against the bare core of development and testing activities, this tempts almost everybody to shortcut the engineering procedures, typically to overcome delays, by postponing to tomorrow things that should be done today. The consequences of these *sins* eventually result in substantial degradations in software quality. The best choices in this respect are to rely on clear and simple software engineering practices, vigorously enforced throughout the life cycle of the switching system.

To clarify the point, we present a brief outline of a software engineering methodology, comprising such steps as:

- Documentation and review of requirements;
- Documentation and review of design;
- Code "walk-throughs";
- Documentation and review of test plan;
- Change control methodology.

The development of big software projects, such as those of switching systems, require the use of *logistical tools* that are often implemented as *software factories* intended to permit any party involved in the software activities to deliver his or her contribution to the whole organization. All the organizations involved in the software activities must use the software factory as the repository in which everybody finds what they need and puts what they have done. This is true for designers, programmers, testing people, and planners alike. The logistical instruments are a natural complement of the organizational practice necessary for the development of switching systems. In no case can they be a substitute for it. Rather they appear as a practical means to enforce operating practices more effectively. Apart from the many variations, a software factory is a network of computing facilities and software tools (running on them), providing in an integrated environment compilers, simulators, debuggers, editors, documentation and clerical work tools, planning tools, and other paraphernalia, as needed in any organization dealing with software activities.

CONCLUSIONS

The choices more commonly found in telephone exchanges about the architecture of their common controls result in the adoption of redundant structures based on the use of synchronous twin units. In these structures, as a basic choice, two processors are used instead of one. As long as everything operates properly, they both execute the same processing activities at the same time, including all the common control application system. Meanwhile, the two twin processors check each other to detect the occurrence of any anomaly, in as timely a manner as possible (within small fractions of a second).

There are two main approaches for the implementation of twin processors: microsynchronous and macrosynchronous controls. In the first case, the two machines execute, step-by-step, the same instruction and check their reciprocal behavior. In the second alternative, the mutual check is carried out at the level of the operating system, at well defined systematic moments, during the dispatching of messages among concurrent application processes, and at the activation of each process instance. The first approach, the more costly, applies to any software and is more efficient. The second exploits properties that are typical of the switching software and consumes substantial amounts of CPU time for the synchronization activities.

In both cases, the two replicas of a common control are provided with internal mechanisms that permit them to detect the occurrence of anomalous situations. These mechanisms are implemented at both software and hardware levels. In the hardware, suitable circuits control the units by analyzing parity conditions and timing constraints on binary signals. In the software, specific programs, part of the operating system, activate checking circuits and verify that every circuit responds as expected in the common control hardware.

Another architecture for redundant common control is based on the use of common data memories built on clusters of CPUs. In each cluster, a set of N_i useful units share the application system, while a further standby unit is available to replace, on-line, any unit that may fail. Unlike the case of dual units, the standby units, when in waiting, do not carry any application system activity. This is possible because they find in common memory all the information that they will need to start their on-line activity, when necessary.

The operating system of a common control virtualizes its machine to appear to the application system as a single never-failing CPU. Therefore, it also includes the software needed for the management of the common control redundancy. The regional processors are supervised by the central control that also checks their behavior by means of routines partially placed in the regional processors, and in the central control proper. The same central control is also capable of switching off any regional processor that it finds to be faulty.

Especially for the case of large exchanges or clusters, support computers also must be implemented as redundant machines. In this case, the use of computer clusters built around data networks offer the possibility of modular growth and the feasibility of loading critical functions on more than one processor to have these functions resume shortly after any CPU failure. The central problems in the case of support computers are related to the availability of mass memories and the integrity of databases in the case of failures. The operating system must therefore include the features needed to offer this integrity.

Improper actions by operators should not generate catastrophic effects in the exchanges. This need demands an intensive and pervasive analysis by the application software of most of the requests coming from operator positions of any kind.

Also, residual faults in the software could have catastrophic effects in the exchange. This circumstance places very stringent demands on the quality of the exchange software. These demands can be met by careful choices in the domain of software architecture programming languages, software engineering, and logistics.

REFERENCES

1. CHILL (Programming Language): "CCITT Blue Books," ITU, Geneva, 1989, Vol. X, Fasc. X6.
2. SDL: "CCITT Blue Books," ITU, Geneva, 1989, Vol. X, Fasc. X2, X3, X4, X5.

INDEX

THE AUTHOR

While earning his degree in Physics at the University of Rome, Giuseppe Fantauzzi was awarded a number of scholarships at the Instituto Superiore delle Poste e Telecommunicazioni, the University of Rome, and at NATO, Paris. Currently Executive Director for Systems at Telesoft S.p.A., Mr. Fantauzzi worked more than twenty years for ITALTEL, where he was first in charge of the R&D Laboratory for automating electromechanical exchanges and developing international digital exchanges, and then the Data Network Laboratory within the Public Electronic Switching Division. He is a member of both the IEEE and AFCET.

The Artech House Telecommunication Library

Vinton G. Cerf, *Series Editor*